SYSTEM VERIFICATION

SYSTEM VERIFICATION

Proving the Design Solution Satisfies the Requirements

Second edition

JEFFREY O. GRADY

AMSTERDAM • BOSTON • HEIDELBERG • LONDON
NEW YORK • OXFORD • PARIS • SAN DIEGO
SAN FRANCISCO • SINGAPORE • SYDNEY • TOKYO
Academic Press is an imprint of Elsevier

Academic Press is an imprint of Elsevier
125 London Wall, London EC2Y 5AS, UK
525 B Street, Suite 1800, San Diego, CA 92101-4495, USA
50 Hampshire Street, 5th Floor, Cambridge, MA 02139, USA
The Boulevard, Langford Lane, Kidlington, Oxford OX5 1GB, UK

Library of Congress Cataloging-in-Publication Data
A catalog record for this book is available from the Library of Congress

British Library Cataloguing in Publication Data
A catalogue record for this book is available from the British Library

ISBN: 978-0-12-804221-2

For information on all Academic Press publications
visit our website at www.elsevier.com

Working together
to grow libraries in
developing countries

www.elsevier.com • www.bookaid.org

Publisher: Joe Hayton
Acquisition Editor: Brian Guerin
Editorial Project Manager: Carrie Bolger
Production Project Manager: Anusha Sambamoorthy
Designer: Matthew Limbert

Typeset by Spi Global, India

CONTENTS

ABOUT THE AUTHOR

Jeffrey O. Grady has been the owner of JOG System Engineering, a system engineering consulting and training company, since 1993, based in San Diego, California. In that capacity he has worked with a wide array of companies with a broad cross section of product lines. Prior to this, following ten years in the U.S. Marines, he worked for 30 years in aerospace companies as a customer-training instructor at Librascope on Anti-Submarine Rocket (ASROC) and Submarine Rocket (SUBROC) underwater fire control systems; field engineer, project engineer, and system engineer at Teledyne Ryan Aeronautical on unmanned day and night as well as strategic and tactical photo reconnaissance, SIGINT, strike, EW, and target aircraft launched from ground launchers, aircraft, and shipboard; system engineer at General Dynamics Convair on advanced cruise missile; and system development department manager at General Dynamics Space Systems Divisions on space transport (Atlas) and energy systems.

Jeff has authored 11 published books and numerous papers in the systems engineering field. He holds a bachelor's degree in mathematics from San Diego State University, a master of science in system management with a certificate in information system from the University of Southern California, and a systems engineering certificate from the University of California San Diego. Jeff teaches system engineering courses around the country on-site at companies and universities including the University of California San Diego. He is an International Council on Systems Engineering (INCOSE) fellow, founder, and expert system engineering professional (ESEP).

PREFACE

The original work for this book began under the title *System Validation and Verification* published by CRC Press in 1997. It was rewritten in 1998 to some extent and coordinated with a rewrite of several other books that I had written to form Volume 4 of a four-volume series titled *Grand Systems Development* for use as text material in a system engineering and development instruction program offered by my consulting and training company, JOG System Engineering. The original book covered both validation and verification but the book was published by Academic Press focused only on verification. The validation material was moved to the companion book in the same series titled *Systems Requirement Analysis* also published by Academic Press.

This is a second Academic Press edition published in 2012. It extends the same process applied to product verification to process verification following the pattern of: define the problem, solve the problem, prove it. Fairly recently, sound standards have evolved for system development, system engineering, and software engineering. These standards act as process specifications for the corresponding processes we must design. Sound standards are also maturing for capability maturity models for use in assessing the quality of the process design providing a process verification mechanism.

What the author recognized while working on a program for a system development enterprise was that he had never observed a clear way implemented on a program to communicate to the person assigned responsibility for a verification task specifically what requirements that person was responsible for producing verification evidence about nor had he observed an effective method applied requiring that person responsible for a task to coordinate the task evidence to those same requirements. That is, a clear means for capturing the traceability between product requirements, verification requirements, verification task, and the content of the report for that task was nowhere described. The author set out to include this information in this book. He also finally concluded in this edition that the integrated verification documentation addressed in the two prior versions had to be replaced by the common separate series of documents for each task most often seen on a program.

Jog
San Diego, California
February 10, 2016

ACKNOWLEDGMENTS

This book remains dedicated to the memory of Mr. Max Wike, a retired naval officer and a system engineer in the department I managed at General Dynamics Space Systems Division in the mid-1980s. Before he passed away long before his time, he and his wife to be, Ms. Debbie Matzek, had succeeded in beginning my education in requirements verification and I very much appreciate that to this day.

It never fails that during every class on system requirements analysis that I have taught for University of California San Diego, UC Irvine, UC Berkeley, through short course companies, and independently through my systems engineering consulting and education firm at companies, students have offered me great insights and ideas about how this process can be done better. Many of these inputs find their way into future courses as well as new books and, hopefully, revisions of others. Unfortunately, I have not shown due diligence in keeping a list of these people by name and location so that I can properly thank them. The list would be very long. By the time this book is published it will likely include on the order of 1000 engineers.

A lady from Raytheon in Tuscon, AZ, Christine Rusch, attending the International Council on Systems Engineering (INCOSE) Symposium V&V tutorial I did in Brighton, England in 1999 suggested coverage of the reverification situation adding content to the course and a chapter to the text. A gentleman attending a similar tutorial at the 2000 INCOSE Symposium in Minneapolis, MN discovered that I had not covered the reporting process adequately, causing the addition of chapters on that subject. Also, at the same time while considering how that material should be added, I noticed that system test and evaluation was covered in a most cursory fashion and a whole part was added to the text. It is true that a lecturer learns a great deal through teaching because you have the opportunity of talking to some very bright people. I appreciate all of those suggestions.

I benefited a great deal from a UC San Diego System Engineering Certificate Program course I attended while the ideas for this book were initially swarming in my mind. That course was titled System Verification taught by a great guy and fine system engineer, Mr. James R. (JB) Hill, at the time a vice president at Scientific Applications International Corporation (SAIC) in San Diego, California. I very much appreciate the new knowledge I derived from his course.

Several people in industry provided much appreciated and valuable insights into modern techniques for validation and verification on items as small as integrated circuits and as large as a transport aircraft. Mr. David Holmes, a system engineer at Interstate Electronics Corporation in Anaheim, California, very generously arranged a demonstration of how they develop application specific integrated circuits (ASIC) during a certificate program I presented for UC Irvine at their facility in 1995. Mr. David B. Leib, a Test Engineer on the McDonnell Douglas C-17 Flight Control System Simulator (Iron Bird), gave me a tour of the simulator at their facility in Long Beach, California during a certificate program I presented at their facility for UC Irvine in 1996 that was very useful in describing a typical large-scale validation and verification instrument.

While I have benefited from the advice of those identified and many others where I cannot recall the precise situation, they should not be held accountable for any errors of commission, omission, or wrong-headed thinking. They belong wholly to the author.

ACRONYMS AND ABBREVIATIONS

Acronym	Meaning
AFSCM	Air Force Systems Command Manual
ASROC	anti-submarine rocket
ASW	anti-submarine warfare
ATP	acceptance test plan or procedure
BIT	business integration team
BUFF	big ugly fat fellow
CAD	computer aided design
CBD	Commerce Business Daily
CCD	capability development document
CDR	critical design review
CDRL	contract data requirements list
CFD	control flow diagram
CMM	capability maturity model
CMMI	capability maturity model integrated
CONOPS	concept of operations
CPM	critical path model
C/SCS	cost/schedule control system
DER	designated engineering representative
DID	data item description
DFD	data flow diagram
DoD	Department of Defense
DODAF	Department of Defense architecture framework
DTC	design to cost
DT&E	development test and evaluation
EFFBD	enhanced functional flow block diagram
EIA	Electronics Industry Association
EIT	enterprise integration team
EMI	electromagnetic interference
FAA	Federal Aviation Administration
FCA	functional configuration audit
FMECA	failure mode effects and criticality analysis
FOT&E	follow-on operational test and evaluation
FRACAS	failure reporting and corrective action system
FRAG	fragment
GD	General Dynamics
HLD	higher level demote
IAT	integration, assembly, and test
ICAM	integrated computer aided manufacturing

Acronym	Meaning
ICBM	intercontinental ballistic missile
ICD	initial capabilities document
ICD	interface control document
IDEF	integrated definition
IEEE	Institute of Electrical and Electronics Engineers
ID	item demote
ID	identification
IMP	integrated master plan
IMS	integrated master schedule
INCOSE	International Council on Systems Engineering
IOC	initial operating capability
IOT&E	interim operational test and evaluation
IP	item promote
IPPT	integrated product and process team
IRAD	independent research and development
ITP	integrated test plan
ISO	International Standards Organization
ITEP	integrated test and evaluation plan
IVDR	integrated verification data report
IVP	integrated verification plan
JCD	joint capability document
LLP	lower level promote
MID	modeling identification
MIL	military
MK	mark
MoD	Ministry of Defence
MODAF	Ministry of Defence architecture framework
MSA	modern structured analysis
NASA	National Aeronautics and Space Administration
OOA	object oriented analysis
ORD	operational requirements document
OT&E	operational test and evaluation
PARA	paragraph
PBT	program business team
PCA	physical configuration audit
PDR	preliminary design review
PDT	product development team
PERT	program evaluation review technique
PID	product identification
PIT	program integration team
PMP	parts, materials, and processes

Acronym	Meaning
PRID	product requirement identification
PSARE	process for system architecture and requirements engineering
PT	part
QA	quality assurance
RAM	reliability, availability, and maintainability
RATO	rocket assisted take off
RPM	revolutions per minute
REQ	requirement
RFP	request for proposal
RID	requirement identification
SADT	structured analysis and design technique
SAR	system architecture report
SAR	surface air rescue
SDR	system design review
SOW	statement of work
SQA	software quality assurance
SRA	system requirements analysis
SRR	system requirements review
STD	standard
STE	special test equipment
STEP	system test and evaluation plan
SysML	system modeling language
TEMP	test and evaluation master plan
TQM	total quality management
UADF	universal architecture description framework
UHL	ultra high frequency
UML	unified modeling language
UPDM	unified process for DODAF/MODAF
USAF	United States Air Force
USN	United States Navy
USS	United States ship
VCRM	verification cross reference matrix
V/H	the ratio of velocity over height above terrain
VRID	verification requirement identification
VTN	verification task number
V&V	validation and verification
WBS	work breakdown structure

GLOSSARY

Chapter 1 Setting the Stage

Compliance A condition of agreement between the content of a document (a specification for a product entity in particular) and the features of that product entity.

Department of Defense Architecture Framework (DODAF) A comprehensive modeling approach developed by Department of Defense (DoD) to be used by contractors in the development of systems for delivery to DoD.

Matrix Management A management structure applied to an organization providing two axes: (1) a set of enterprise functional departments, each one of which is focused on a particular kind of work under a functional department manager, and (2) a set of programs each organized under a program manager into a set of cross-functional teams, each focused on an important product entity within the system the program is developing.

Ministry of Defence Architecture Framework (MODAF) A comprehensive modeling approach adopted by the British Ministry of Defence based on DODAF.

Principal Engineer A person specifically assigned responsibility for the accomplishment of a task or activity.

Program Classes All programs are partitioned for management purposes into one of three classes: (1) high-rate production programs where many product articles are produced over a considerable period of time, (2) low-volume, high-dollar production programs where only a few product entities are produced but each has a high cost, and (3) one-of-a-kind programs where only a single system is produced and delivered.

Specialist An engineer tightly and deeply focused and qualified in a narrow field of knowledge important in the business in which the enterprise is engaged.

Specification A document containing the essential characteristics of the design of a product entity.

System A collection of product entities and other entities that collectively can be applied to achieve a desired outcome.

System Architecture Models Four UADF are recognized: (1) functional model employing functional flow diagramming and a set of other models, (2) the combination of modern structure analysis (MSA) and process for system and requirements engineering (PSARE), (3) the combination of unified modeling language (UML) and system modeling language (SysML), and (4) unified process for DODAF and MODAF (UPDM) implemented in SysML. The intent is that an enterprise would select one UADF and apply it on all programs.

System Development An organized process defined in a plan, budget, and schedule to achieve a specific goal measured in terms of a specific list of achievements by a specific date in relation to an intention to deliver a system to a customer in accordance with a contract.

System Engineer An engineer with a broad interest in the whole rather than the parts possessing a gift in encouraging good communication between specialists from different functional departments or knowledge domains.

Universal Architecture Description Framework (UADF) A modeling capability useful in defining the architecture for a system and its included entities and developing the content of all program specifications. The model must be effective regardless how the product design will be implemented in terms of hardware, software, and people doing things.

Validation The process by which a person or team of persons determine the degree of compliance between a product entity and the content of a customer prepared document describing or defining needed mission capabilities for a product entity.

Verification The process by which a person or team of persons determine the degree of compliance between a product entity and the content of the specification for that product entity.

Verification Classes All verification work is partitioned for management purposes into one of X classes: (1) system test and evaluation either accomplished by the contractor as a development test and evaluation (DT&E) process relative to the content of the system specification or as a customer-implemented operational test and evaluation (IOT&E) process relative to the content of a customer prepared mission-oriented document, (2) item or interface qualification verification to determine the extent to which the design of the entity complies with the content of the item or interface performance specification and is qualified for the intended application, (3) item or interface acceptance verification to determine the extent to which the item or interface is acceptable for delivery to the customer as defined in the item or interface detail specification, and (4) parts, materials, and processes verification for items identified on engineering drawings for items and interfaces to determine the extent to which those parts, materials, and processes comply with the content of supplier specifications and other documentation.

Verification Levels All verification work is partitioned for management purposes into one of six levels: (1) system, (2) item, (3) interface, (4) part, (5) material, or (6) process.

Verification Methods All verification work is partitioned for management purposes into one of six specific methods: (1) analysis, (2) test, (3) demonstration, (4) examination, (5) special, and (6) none. Special is associated with the use of modeling and simulation.

Chapter 2 Specifications and Their Content

Detail Specification A specification that prescribes the features with which a product must comply in order to be acceptable for delivery.

Performance Specification A specification that prescribes the features of a product in terms of required performance characteristics with which the design must comply. All system specifications are performance specifications.

Chapter 3 Specification Section 3 Preparation

Boilerplate Slang term for generic content of a set of specifications.

Environment All of the entities and relationships not a part of a system that have a relationship with that system.

Flowdown Content of a parent specification believed to be appropriate for inclusion in a child specification.

Functional Analysis A modeling technique employing functional flow diagrams to analyze what a system must accomplish and how it must perform. Originally applied to hardware situations before software was invented but subsequently adapted to software development.

Modern Structured Analysis (MSA) A modeling technique employing bubble diagrams with connecting lines reflecting data transfers between functional bubbles. Developed for software development.

Process for Architecture and Requirements Engineering (PSARE) A modeling technique extended from MSA by Hatley and Pirbhai, at one time referred to as the HP method, that can be applied to systems, hardware, and software development.

Chapter 4 Specification Section 4 Preparation

Analysis Verification Method Product item features are studied employing technical or mathematical thought processes to provide evidence of the degree of compliance the design achieved relative to the content of the item specification.

Demonstration Verification Method Product items are subjected to actual operation, adjustment, or reconfiguration in accordance with approved procedures to expose the degree to which the item characteristics comply with the content of item specification.

Examination Verification Method A product item is inspected without the use of special laboratory appliances or procedures to determine the extent to which the product features comply with the content of the item specification.

Special Verification Method A combination of other verification methods or the use of simulation and modeling to determine the extent to which a product complies with the content of the item specification.

Test Verification Method The application of technical means often involving the use of special equipment and procedures to determine the extent to which the features of the design of an item comply with the content of the item specification.

Verification Requirement Content of a specification that prescribes the essential characteristics of the verification work that must be accomplished to determine the degree to which a product complies with the product requirements included in that specification.

Chapter 5 Establishing and Maintaining Enterprise Readiness for Program Implementation of Verification

Requirements and Verification Database/System A computer application designed to support the development of specifications and accomplish related verification work.

Specification Data Item Description (DID) A document that tells how to transform a specification template into a specification.

Specification Template A preferred or prescribed paragraphing structure for a specification that is to be employed on all programs within an enterprise.

Chapter 6 Verification Process Design for a Program

Development Test and Evaluation (DET) Testing accomplished during design work to evaluate design concepts and alternate approaches to design problems.

Chapter 7 Parts, Materials, and Processes Verification

Long Lead List A list of parts and materials that will be required to manufacture the items that will be consumed in item qualification verification for the purpose of obtaining customer approval for reimbursement of money spent to acquire these parts and materials long before they will be required to support normal item manufacturing.

Chapter 8 Item and Interface Qualification Verification

Functional Configuration Audit (FCA) A formal review of the qualification verification evidence for an item by the customer. Approval commonly leads to customer approval for production.

Personal Integrity Successful verification work must be accomplished by people with integrity who will report the results of verification work truthfully.

Qualification Verification Verification work applied to product and interface entities for the purpose of determining the extent to which the item or interface design complies with the content of the performance specification.

Verification Cross Reference Matrix (VCRM) A tabular display of all requirements appearing in program specifications coordinated with all related verification information. Also referred to as a compliance matrix.

Chapter 9 Item and Interface Acceptance Verification

Acceptance Verification Work accomplished to determine if a manufactured article complies with the content of the item detail specification so as to be approved for delivery to the customer.

Physical Configuration Audit (PCA) A formal review of the acceptance verification evidence for an item by the customer.

Chapter 10 System Test and Evaluation Verification

Certification A process required in some product fields to ensure that a design complies with the content of a document maintained by an organization in accordance with some official status.

Development Test and Evaluation (DT&E) Verification work accomplished at the system level by the developer to determine the degree to which the system complies with the content of the system specification.

Operational Test and Evaluation (OT&E) Verification work accomplished by the user to determine the extent to which the system will comply with the content of user documentation defining mission requirements.

Chapter 11 Enterprise Process Verification

Program and Functional Metrics Measurements from program performance supporting effective management of program activity.

Cost/Schedule Control System Commonly a computer implemented system for controlling program cost and schedule in support of program management.

Chapter 12 System Test and Evaluation Verification

Grand Planning Environment A multi-faceted view of program management.

CHAPTER 1

Setting the Stage

1.1 THE ENDURING TRUTH THAT NEED NOT BE

It is a sad truth that many system development programs fail to deliver a system to the customer with understandable evidence of system compliance with the content of program specifications. In order to be successful in doing so, the developer must begin early in the program to develop a good set of specifications in which the content has been derived through the application of an effective modeling method accomplished by people who know what they are doing. Then, the content of the specifications must be applied as the basis for design of the product. This same content of the specifications must also be used as the basis for the design of a verification process that produces evidence in a comprehensive series of verification task reports from which any reader will reach the same understanding of the degree of requirements compliance achieved. This is the third book on this subject the author has had published and it is the first time he has included the details about how to accomplish verification successfully and affordably, based on some recent program experiences in which a correction for the sad story became obvious to him but was not possible to achieve, because the driving specifications were not well-crafted by the customer and contractor and the remaining program budget, schedule, and customer goodwill were in insufficient supply to support correction of past errors. It should be said, however, that despite this past history the product system delivered was superb. The point of this book is that it is possible to deliver both a great product along with understandable verification evidence that truthfully reveals that the delivered product complies with the content of the program specifications.

The road to success is simple but it appears not to be easy to accomplish. In this book we will identify a simple prescription that any program can follow to succeed in system verification, but you must begin the prescription fairly early in the program and management must follow the plan through with determination and management skill.

System Verification
http://dx.doi.org/10.1016/B978-0-12-804221-2.00001-2
1

1.2 OVERVIEW OF THIS CHAPTER

This chapter collects in Section 1.3 general information about verification, intended to prepare the reader with background information that it is hoped will help the reader place the information provided in the other chapters in context. Section 1.4 continues the overview of the remaining chapters of the book.

We begin the introductory ideas in Section 1.3.1.1 with some fundamentals about the verification process definition, scope, and goals. In Section 1.3.1.2 we identify the customer relationship and the important role it plays in the verification process whether the sale is through a contract or a commercial sale. In Section 1.3.1.3 we emphasize the importance the words *truth* and *integrity* play in the verification process.

Section 1.3.1.4 correlates specific sets of specifications with particular verification classes and provides a checklist of specific actions that are necessary related to these specifications and classes. Section 1.3.1.5 discusses the partitioning of classes into product entities that focus our verification work, which is further partitioned into tasks of six kinds.

In Section 1.3.1.6 the meanings of two important words starting with the letter "v" are covered as these words are used in the book. Finally, a foundation of systems engineering is provided based on the limited capacity of the individual human mind relative to the amount of information that mankind has accumulated. This will lead to the importance of human communication to link the minds of several (maybe hundreds of) human beings who have each focused their knowledge capacity on a narrow field deeply. It will also show the need for people called *system engineers*. The management challenge is to form the equivalent of one great mind that has the capability to solve problems that no single person can possibly deal with successfully. Hopefully, this discussion will reveal to the reader why management is the most difficult task on a program and why all of us who are managed should do our best to support managers trying to do their best in a very difficult activity.

Section 1.3.1.7 offers ideas about attaching responsibility for the verification work to program personnel. This can be a complicated relationship on a program but the fundamentals are that one person should be identified as responsible for every verification task and these tasks should be managed through a well-defined management structure. A particular structure is offered but there are many reasonable ways to set this up on a program. Section 1.3.1.8 notes a need to assemble all of the parts of the verification process into an overall program plan that is further discussed in detail in Chapter 6.

Section 1.3.2 defines a system and provides an overview of the systems development process focused on the three fundamental phases of a program: (1) define the problem in a set of specifications; (2) solve the problem through synthesis of the requirements contained in the specifications in the three steps of design, procurement, and manufacture; and (3) determine the extent to which the resulting product complies with the content of the specifications. This book focuses on the third phase, verification. A fourth important area of interest is to provide good management across the complete program and within the enterprise across its active programs. The case is made for an enterprise possessing a single common process applied to all programs and covers four development environments within which an enterprise can apply this common process on programs: (1) waterfall, (2) V, (3) spiral, and (4) agile. In all cases the three-step development focus overlaid by good management is applied, being completed in the verification phase that actually extends throughout production, however long that may last. A program can continue far beyond its initial production phase of course, as in the case of the Boeing B-52 program, and very often parts of the four verification phases may have to be implemented relative to major modifications.

Section 1.3.2 also summarizes the work that should precede the synthesis and verification work to prepare a set of specifications deriving the content of those specifications through the application of a comprehensive modeling approach that the enterprise has prepared its engineers to employ. Four universal architecture description frameworks (UADF) are offered from which an enterprise can select to be applied on every program, no matter how it is determined that the problem should be solved through a design implemented in some combination of hardware, software, and people doing things. The reader is encouraged to consult the author's book titled *System Requirements Analysis,* Elsevier, 2014 for full details on the methods summarized in this section.

Section 1.3.3 provides a description of the author's preferred organizational structure that is assumed throughout the book, involving a matrix management structure. In this structure a set of functional departments, each specializing in a particular discipline, provides programs with the resources they need to accomplish planned work. The program organizational structures, each under a program manager, then organize these resources into cross-functional teams oriented around the product entities the program system will require.

Section 1.3.4 covers program phasing possibilities, describing three specific phasing concepts. The application of these phasing approaches tends to be fairly rigid in practice, and on some programs where it is very difficult to define the requirements comprehensively in a timely way program managers have applied what is called *rapid prototyping* or *agile development* in an attempt to speed up the process and avoid late recognition of serious problems, in some cases with success. Finally, the section covers three different program cases in terms of the number of systems delivered.

1.3 INTRODUCTORY IDEAS INCLUDED IN CHAPTER 1

This section offers important introductory ideas referred to in the preceding paragraphs, while Section 1.4 provides an overview of the other chapters of the book.

1.3.1 Our Verification Objective

1.3.1.1 What Is Important?

This book is titled *System Verification*. Therefore, the reader would expect that it concerns establishing the truth or correctness of a proposition and that expectation will be fulfilled. But what do we seek to verify and how shall we go about it? The assertion we will deal with is primarily of the kind, "product entity X complies with content Y of document Z" where: (1) the entity X is a product design and/or manufacture for a system or part thereof; (2) content Y may be restricted to a paragraph, a collection of paragraphs, or the content of a complete specification; (3) the document in question Y is a specification for that entity; and (4) compliance means that the design and/or manufacture of the entity is in complete agreement with the content of the document, often phrased as being in compliance with that content.

We should recognize that a product entity has many representations including the physical entity itself, a collection of engineering drawings and corresponding parts lists, a collection of analysis and test reports, a specification, procurement documentation, and manufacturing documentation. It is insufficient to be able only to say that the physical product complies with the specification content. We need all of the representations to also agree one with the other but we focus most of the energy of verification on the physical product entity that will be delivered relative to the content of its specification on the theory that any noncompliance contained in the supporting documentation should result in a failure of the product to successfully complete verification. It would still be possible, of course, that one error in supporting documentation in combination with an error in verification planning could produce an inappropriate compliance conclusion, but that occurrence would have a low probability.

Verification occurs in the systems development process at the tail end of a trio of processes begun by requirements and synthesis. In the systems development process we first must determine what the system will consist of in terms of product entities and relationships between them through a process some, including the author, like to refer to as *architecture and requirements engineering*, concluding with what the essential characteristics of all of those entities must be. We must then accomplish synthesis of the requirements for any of those entities to create a physical reality representing the product entity through design, acquisition or procurement of the resources needed to produce or manufacture the entity using those resources. The apparatus within which the manufacturing work takes place can be as complicated as the product entity being manufactured (or more so) and may require the same development process that is applied to the product design – an automobile or commercial aircraft final assembly plant are two cases in point.

The product entity in question may be software, instructions for human beings to perform in a prescribed fashion toward a particular end, or a hardware product. In some cases it will be possible to reach a proper conclusion about compliance without possessing an example of the product entity, but in most cases we will find it necessary to inspect an article of the product entity in some detail. The word *inspection* in this case is commonly applied as embracing five different methods: (1) analysis of representations of the product entity such as drawings or data; (2) test of the product entity using special test equipment; (3) examination of the product entity using human senses and simple devices; (4) demonstration; and (5) special, which is reserved in this book to apply to the use of modeling and simulation. We will tag on a "none" method, meaning that no verification is necessary. This method is appropriate in the case where a paragraph in a specification includes no requirement. An organizing title-only paragraph is an example of the latter. In all cases these methods are applied with the expectation that the product entity can accomplish some particular activity under specific conditions and within a specific time frame.

For a product entity that follows this path through a well-managed system development process staged by a program operating within an enterprise in accordance with a well-designed plan, we may say that the development for that entity has been successful and articles of that product entity can be delivered to a customer where they are operated and maintained in accordance with instructions agreed upon between developer and customer. The word *customer* used in this book may include a pair of organizations in cases where one agency of the customer's organization acquires the system under a contract between that acquisition agent and the developing enterprise and another uses it during its life cycle. Most systems developed for the U.S. Department of Defense are examples.

The important distinction for the reader to take away from this discussion is the need to preserve full life expectancy for every manufactured article delivered to the customer while ensuring that each will satisfy the requirements in the related specifications.

1.3.1.2 Customer Relationship

It is possible to partition most system development situations into: (1) those that take place in accordance with a contract between a customer and a developer where the developing enterprise delivers the system in accordance with the contract; and (2) those that take place within an enterprise for commercial sale to a customer base either through direct sale or some form of retail distribution. In either case the developer should seek to verify that the product entities delivered comply with the requirements defined as a precursor to design. In the first case, it may be a part of the contract that the developer must supply the customer with verification reports. In the second case, access to the requirements respected and proof of requirements compliance are not commonly consciously expected by the customers but, in the author's view, a developer with integrity would include verification cost and schedule impact in the development of its product.

Personnel in responsible positions of commercial developers may very well have the attitude that verification evidence includes competition-sensitive information that should not be shared with competitors. However, in some commercial product lines, generally involving some form of public safety, government agencies regulate the development and manufacture of the product requiring proof of compliance with those regulations and intimate knowledge of product features. Federal Aviation Administration (FAA) influence over aircraft production is an example. The Food and Drug Administration has a similar interest in the products they regulate.

This book is primarily focused on the case in which a developer delivers a product in accordance with a contract, because the author's primary experience has been involved with development of systems for delivery to large customers like Department of Defense (DoD) and National Aeronautics and Space Administration (NASA). However, the content of this book applies as well to the development of automobiles, computer software, medical equipment, and toothpaste.

1.3.1.3 Our Scope of Interest

So, verification is a process for determining the truth about the design and/or manufacture of a product entity relative to the content of its specification. Truth is a very important element of this word and process. If a product fails to comply with the content of its specification yet its developer reports to its customer that it does comply with the content of the specification, the developer lies and should not be rewarded in accordance with the contract that presupposes that delivered product will comply. Those who would do verification work for an enterprise must recognize that they have a responsibility that goes beyond the boundary of their employer, leading to potential conflict between employees who do this work and their managers. Managers more interested in delivery schedules and financial statements have been known to encourage shipment no matter the product performance. The customer has every right to expect that all verification work the contractor accomplishes and any reports it may prepare about that work will be characterized by integrity.

In each verification activity, however slightly or grandly we are making a comparison between a product entity and the content of a specification, the content of the specification is a very important part of this process of comparison. It is possible for the customer or developer to prepare a poor specification and for the developer to accomplish the related verification planning and procedural documentation poorly, such that when the verification work is complete it could be incorrectly concluded that the product design complies with the content of the specification. So, in order for a developer to qualify for a mythical badge of product development integrity, the documentation as well as the design must be in compliance with some very high standards.

This book does explain how to determine what verification work must be done, how the total task can be broken down into some number of verification tasks involving the six methods (including none) mentioned earlier, how to prepare a plan, procedure, and report for each of these tasks, and conduct an audit of the content of those reports for a particular product entity. The author's companion book *System Requirements Analysis*, also from Elsevier, provides full coverage of how to prepare an effective set of specifications for a program, which can be used as a basis for all program synthesis and verification work. The cautious manager may conclude that the process of performing all of the work encouraged in these two books will squeeze out any potential for profit, but there is a great truth in systems development that is sometimes stated as "pay me now or pay me later." This flippant remark, uttered by a system engineer to a proposal manager when he or she is in the process of significantly reducing the budget proposal for requirements analysis or verification planning work, means that if you reduce the early budget for important systems engineering work you will have to pay considerably more to work the program out of a serious problem at a time when all of the money, schedule slack, and customer goodwill have been spent. Some managers learn this after one unfortunate mistake and some never do. The performance of the work described in these two books exposes an affordable systems development approach, when accomplished and managed by people who know what they are doing. Hopefully, the content of these two books will also be helpful to functional department managers in determining how to supply enterprise programs with the resources they need to attain success. Cooperative work between functional department managers over time can provide the enterprise with a common documented process and supporting resources, applied on all programs and incrementally improved, that result in a continuing training program of no cost through common process repetition and incremental improvement.

1.3.1.4 Verification Phasing

The verification work required on a large program calling for the development of a number of new product entities at several levels of indenture will involve quite a bit of complexity and will be stretched over a considerable period of time that begins before all of the specifications have been completed, runs through much of the design and procurement work and completely through the manufacturing work on the program, and continues so long as manufacturing work carries on. So, do not think of the verification work as being some minor program activity, the pain of which will pass in a little while. Many experienced managers and engineers who have worked on verification on a program that did a poor job of preparing its specifications – leading to design engineers not being able to clearly understand how to comply with the requirements in their design and verification task principal engineers having trouble determining whether or not the design complied – have equated the words *verification* and *pain*. It need not be this way. If a program manager wants an affordable and successful verification experience, such an experience must be preceded by:

a. Availability of a reviewed and approved system specification defining all essential system characteristics derived through an application of an effective modeling approach wherein modeling work products (commonly sketches) and traceability between the modeling artifacts on those work products and the requirements derived from them were captured and configuration managed along with the system specification content. The specification is to include verification requirements prepared in each case by the person who prepared the corresponding product requirements and a table that reveals traceability between product requirements and verification requirements, the method that is to be applied in verifying (test, analysis, demonstration, examination, special, or none), and the level at which the verification work will take place (child, item, or parent).

b. Availability of a set of item performance specifications for all product entities in the system that will require development, with the content in all cases derived as described under the system specification using the same modeling approach. All of these specifications and related modeling work products reviewed, approved, and placed under configuration control.

c. Availability of a set of item detail specifications, each corresponding with an item performance specification, containing requirements that must be satisfied by every product article before shipment to the customer. All of these specifications reviewed, approved, and placed under configuration control.

d. Capture of appropriate documentation (including specifications) on all parts, materials, and processes (PMP) listed on engineering drawings for system entities in accordance with an effective program standard PMP activity that encourages use of proven material in the design of those entities.

e. Product design, procurement, and manufacturing planning accomplished by cross-functional teams staffed from appropriate engineering and support disciplines working under the management of a team leader responsible for team item cost, schedule, product features, support for related procurement actions and manufacturing planning. All of these teams served by system engineers who are skilled in communications, integration, and optimization across product and domain boundaries. All design, procurement, and manufacturing documentation placed under configuration control.

f. Product preliminary and detail design reviews conducted for each product entity at some level of team indenture with published minutes placed under configuration control.

g. Physical co-location of team personnel and encouragement of individual and team interaction in conversation about development of the product for which the team is responsible. Team work products should be formally reviewed and approved followed by capture of master copies by configuration management such that any changes require formal review and approval.

The verification work is commonly partitioned into four classes, each of which is linked to a set of specifications as follows:

a. Parts, materials, and processes (PMP) identified on the face of engineering drawings or in related lists must be determined to be compliant with requirements supportive of the items within which they are included. The principal element of a program PMP activity is the development and maintenance of an approved PMP list that identifies PMP proven to be adequate through prior use or verification of the content of a supplier PMP specification.

b. Item and interface qualification verification is verification work conducted at the item level for the purpose of determining the extent to which the item is qualified for the intended application. An item that passes this inspection process is said to be qualified and to have completed qualification for the intended application. The planning related to item qualification verification is driven by the content of the item performance specifications. The union of all qualification task reports for an item must provide evidence that the product design complies with the content of its performance specification. Qualification is normally accomplished only once on a product entity. Generally the items that compose end items of the system must have completed qualification prior to the beginning of development test and evaluation (DT&E) in the interest of the safety of those who will conduct the system testing. Specification and/or design changes subsequent to completion of qualification will commonly require some form of delta qualification to ensure that qualification status remains effective.

c. System test and evaluation is verification work conducted at the system level, in some cases involving isolated end items and the planning for this work is driven by the content of the system specification. All subordinate product entities entering this phase must have been previously qualified for the application. This book recognizes two phases of system verification. The first is often referred to as development test and evaluation (DT&E), conducted by the contractor to determine the extent to which the system complies with the content of the system specification. The second is often referred to as operational test and evaluation (OT&E), which is conducted by the user to determine the extent to which the delivered system complies with the user requirements. A lot will be said later about the meaning of the v words relative to these two kinds of verification work at the system level. Considerable risk can enter a program when the system acquisition customer approves a system specification that is in conflict with a user requirements document. These differences should be dealt with at the system requirements review (SRR) and system design review (SDR) and any deviation on the customer's part (procurement or user agencies) regarding the conclusions reached about the meaning of requirements between these reviews and any later point in the program must be immediately discussed with the acquisition customer. Problems in OT&E often have their beginnings early in a program due to a misunderstanding at a major review or in message traffic during the program.

d. Item and interface acceptance verification is verification work accomplished at the item level for the purpose of determining the extent to which a manufactured article of the item is acceptable for delivery to the customer. The planning related to item acceptance verification is driven by the content of the item detail specification. It is commonly accomplished on every article during and/or after manufacture.

Put these four classes together in time and one observes the complete system verification process. Note that system test and evaluation is but one class in the overall process and deals with system level activity. While it often includes the word *test* in its title, system test and evaluation can and commonly does also include tasks employing the other methods.

Item and interface qualification is performed for items that have not been previously qualified for the application. In some cases, an item design may have evolved or been modified from a previous design or an existing design is being applied to an application that must endure significantly higher stress than that for which the design had been previously qualified; in these cases, a qualification process may have to be applied only relative to the changes or to the extent stress has increased. A developer may also attempt to gain customer acceptance of an item having been qualification verified through what they call "similarity," meaning that the design and application are similar to a prior qualification action. In these cases, there should be an agreement that the developer provide an analysis of the two situations showing the precise degree of similarity and the areas where significant differences exist. The program in this case should have to produce verification evidence of compliance with requirements in all cases where similarity cannot be applied.

Item and interface qualification commonly takes place early in a program before a proper manufacturing process has been set up, so if the program will require the availability of a physical product entity to subject to test, demonstration, and examination, the articles required to support qualification will have to be manufactured by the developer or its supplier using an engineering shop. The program should, in these cases, prepare preliminary manufacturing and acceptance plans and procedures in time for them to be applied in this early manufacturing process. This is important because the customer will understand that the qualification articles will have been manufactured by highly qualified engineers and technicians who will probably not be representative of the personnel who will eventually be present on the full-rate production line for the system. The developer must prepare itself for the customer question, "Why should I believe that the qualification articles produced in a laboratory are truly representative of those that will later flow from the production line?" A reasonable response to that question is that developer personnel followed preliminary manufacturing instructions and acceptance verification procedures while in the process validating those documents for later use in the full-rate production process.

Item and interface acceptance will commonly take place on the production line and involve examinations accomplished by humans applying their normal senses, perhaps aided by simple instruments as well as functional testing. But this verification work differs significantly from that applied in qualification. When acceptance is completed for a product entity in the production environment, that entity must possess full life. The acceptance verification actions must not overstress the article. Commonly, when an article completes qualification it does not possess a full life expectation and in fact may not function at all because it may have been stressed to find out how much margin the design possesses in one or more parameters, such as structural integrity. Items completing qualification may have other program applications, however, even if they no longer function. An inoperative guidance set could be employed in training technicians to disassemble, assemble, remove, and install the unit, for example.

1.3.1.5 Verification Class Composition

Each of four verification classes (system test and evaluation, item and interface qualification, item and interface acceptance, and PMP) comprising the overall verification process should be partitioned into some number of product entities that will be subjected to verification work and the verification work for each of these product entities partitioned into some number of verification tasks. A preliminary partitioning can be made for item qualification and item acceptance relative to the items that will require verification. The result is a specific list of product entities including the whole system and some number of lower tier product entities, not necessarily every item composing the system. Figure 1 illustrates the product entity structure for a system down to the third level and it has been superimposed with shading for

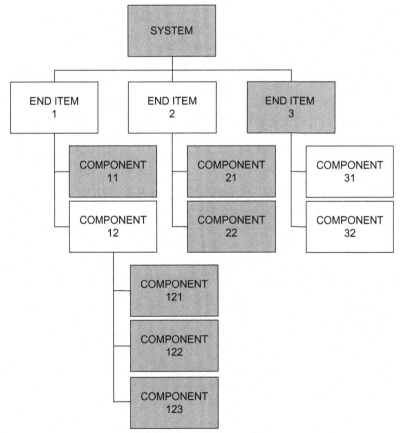

Figure 1 System items requiring verification example.

the entities that we have determined will have to be subjected to system verification and/or item qualification verification, based on an evaluation of many factors that will be discussed in the body of the book. For any given system this selection will be situational.

The fact that the blocks for End Items 1 and 2 are not shaded does not mean that it will not be determined whether End Items 1 and 2 are design compliant. It means that some compliance may be demoted to the child level and in other cases will be promoted to the system level in system test and evaluation, whereas End Item 3 compliance will be determined at the item level.

For each item that will have qualification performed at that item level, the work will be partitioned into some number of tasks of the six methods identified earlier. Each of these tasks will require identification of a principal engineer who will be held responsible for preparing a trio of documents (plan, procedure, and report) and preparing for and accomplishing the work. The plan and procedure in each case will identify a set of task activities that must be accomplished in a particular sequence; the report will present the results in each case, motivating a conclusion about the degree of compliance the product exhibits relative to the requirements in the specification by which the verification task is driven. This exposes the complete organization of the verification work on a program: (1) four classes of verification; (2) accomplishment of item qualification and item acceptance on some number of items (somewhere between 100% and something less than 100%); and (3) for each entity (system or lower tier item) subjected to verification, the development of a trio of verification documents for each verification task accomplished employing one of six methods (test, analysis, demonstration, examination, special, and none). This book calls for separate plans and procedures, but some organizations may prefer to combine the plan and procedure into a document simply called "the plan" with no argument from the author. The PMP verification work will take on a somewhat different pattern, but it will be focused on specific entities covered in each case by a specification as well.

1.3.1.6 The Two Vs

The reader is forewarned that the meanings of two words beginning with the letter V are not universally agreed upon. There are at least three different meanings attached to these two words. These words are *validation* and *verification* (V&V). The dictionary is not much help in explaining any difference between them, and we humans find many other ways to differ on their meanings.

The word *validation* will be used in this book to mean a process carried out to demonstrate that one or more requirements are clearly understood and that it is believed to be possible to satisfy them through design work within the current technological state of the art, funding, and schedule based on demonstrable facts. This is a part of the process of challenging the need for particular requirements and specific values tagged to those requirements prior to the development of solutions or as part of a process of developing alternative concepts, one of which will, hopefully, be the optimum solution in satisfying the requirements. Many system engineers prefer to express the meaning of validation as proving that the right system is being or has been built. The author truncates this meaning to accept only the first part of it – that is, proving that the right system is being or is going to be built – and refers to all post design V&V work as verification that requirements have been complied with.

The word *verification* will be used to mean a proof process for unequivocally revealing the relationship between the content of a specification and the characteristics of the corresponding product. Verification work is done after the design process. It is not correct to say that verification is a process to prove that the product design satisfies its requirements, because this statement implies that, even with a failure of the evidence to show compliance, we might seek to prove compliance anyway. This work requires a high degree of integrity on the part of those who do it because the objective is truth rather than a predetermined outcome. A shorthand statement of the meaning of verification is commonly voiced as, "Did I build the system right?"

The word *verification* was defined in MIL-STD-2167A, a military software specification standard no longer maintained, as a process of ensuring that the information developed in phase N of a development process was traceable to information in phase N-1. Validation, in this standard, was the process of proving compliance with customer requirements.

Validation, as used in this book, therefore precedes final approval of the design solution, and verification follows the design in the normal flow of events in the system development process. It should be pointed out that these words are used in an almost exactly opposite way in some circles, as noted above. The professional system engineer and program manager must be very careful to understand early in a program the way their customer or contractor counterparts are using these two words to avoid later confusion and unhappiness. People from different departments in the same company may even have different views initially on the V words, and these should be sorted out before the company consciously seeks to standardize its process and vocabulary.

These V words are also used in technical data development in the Department of Defense community. The contractor *validates* technical data by performance of the steps in the manual, commonly at the company facility using company personnel. After corrections for errors observed, maintenance and operations technical data is then *verified* by the customer, generally at the customer's facility or area of operations, by demonstrating accomplishment of content using customer personnel, frequently after some training has been completed. When the author was a field engineer for Teledyne Ryan Aeronautical, a gifted technical data supervisor and former Navy Chief Warrant Officer, Mr. Beaumont, informed him, in a way that has not yet been forgotten, how to remember the difference between validation and verification of technical data. He said, "A comes before E! Now, even a former Marine can remember that, Grady!" Having long ago recovered from that Navy slam, the author has come to accept that Mr. Beaumont's simple statement also applies to requirements validation and verification in his book.

The system engineering standard EIA 632, initially released by the Electronic Industry Association in 1999, includes an addition to the definitions of these words. It defines verification essentially as the word is used in this book, except that it breaks off a part of the process related to system verification and refers to it as product validation. EIA 632 defines the term *requirements validation* similarly to the way it is used in this book and offers a process called *product validation* that is intended to prove that the product satisfies its user requirements. The author would put this latter element under the heading of verification work at the system level, verifying that the system satisfies the requirements contained in the original or updated user requirements documentation. In DoD work, this process is called operational test and evaluation (OT&E). The problem with the EIA 632 definition of product validation is that it encourages the notion that there can be valid requirements at the system level external to the system specification that were not part of the development contract.

This can be especially troublesome in a DoD development, where the user community employs a preliminary requirements document, operational requirements document (ORD), concept of operations (CONOPS), or initial capabilities document (ICD) to capture their mission-oriented requirements. All of the content of this user document may not have been accepted into the system specification placed on contract by the acquisition agent, based on a lack of the necessary funding to procure a system that can satisfy all of the requirements in the user document. If the user community verifies against these requirements during OT&E and some of them are noncontractual, the system may not be found acceptable even though it passed all system verification tests conducted by the developer during DT&E, based on the system specification content that was the basis for the development contract. The system requirements review (SRR) and system design review (SDR) are good times to have a conversation between contractor and customer about these matters with capture of the conclusions in the minutes. We will come back to this point later under system verification.

ISO/IEC 15288 offers meanings of these words essentially in agreement with EIA 632, continuing the potential problem with user requirements that did not get picked up by the acquisition agent in the system specification included in the contract.

1.3.1.7 Who Should Be Responsible?

Verification is a very important part of any system development program and someone in a program organization should be held responsible for all of its parts. First, at the enterprise level we should accept that it is primarily an engineering problem while also recognizing that the item acceptance phase will be accomplished within the manufacturing area at a time in a program when manufacturing influence is strong compared to the engineering influence. The work of this person early in a program will be fairly light, growing more time consuming and difficult as program specifications are developed, and becoming very difficult as item qualification planning is entered. This person should interact cooperatively (dashed line) with a collection of people as shown in Figure 2, occupying the following positions: (1) lead system test and evaluation verification engineer, (2) lead item qualification verification engineer, (3) lead item acceptance verification engineer, and (4) lead PMP verification engineer. Management responsibility should come from the program manager and chief engineer and flow through item team managers.

If system test and evaluation is organized into several sites where work will be conducted, a person should be selected to lead the work at each site. Item qualification verification should be organized under the qualification lead into item principal engineers and each task under each item should come under a task principal engineer. Acceptance can be organized in a similar fashion. Parts, materials, and processes verification work is largely clerical in nature because each part, material, or process should be selected by the design engineer from a standard PMP list. Where the design engineer finds no item listed that is adequate, then the design engineer must request that a

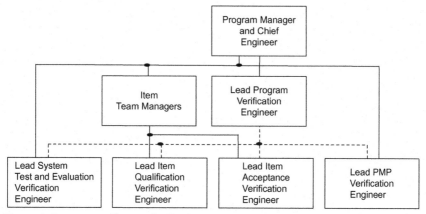

Figure 2 Program verification leadership.

new item be added that the engineer believes will satisfy the need. Then the lead PMP engineer must review supplier documentation and discuss compliance with the requesting engineer. If all of the work at one site is related to a single activity, then this person can also lead all of the tasks through task principal engineers.

The author is a believer in cross-functional teams and encourages that programs be organized with team leaders assigned responsibility in some pattern for the development of product entities comprising the system, reporting up to a chief engineer and program manager. Most or all of the item qualification work will take place during a program period of time when these teams will be staffed, and the author would encourage that all item team lead engineers and verification task lead engineers for those items be selected from the item team. Figure 2 shows only one block for team managers, but there would commonly be several item teams, each led by a principal engineer. There could even be two or three levels of these teams depending on the size of the product system. This diagram is based on the assumption that the Chief Engineer would be the "system team" manager. The responsibility for PMP would be distributed across all of the teams but a program could assign a lead engineer to run the preferred PMP list and coordinate other PMP responsibilities, covered in Chapter 7.

Each verification task will be one of five kinds (test, analysis, examination, demonstration, or special), recognizing that the "none" class will involve no verification work. The analysis task principal engineers will commonly be able to accomplish the task in a completely unaided fashion at their workstation. A principal engineer responsible for a test, demonstration, or examination task may have to depend on a laboratory operated under the responsibility of a Test and Evaluation department. An examination task for a small valve control unit could be accomplished on the desk of the principal engineer but an examination task accomplished on a locomotive diesel engine would more likely have to occur in a work area containing an overhead hoist, railroad tracks, and considerable space. Test and demonstration tasks will commonly also require laboratory space and supporting resources. The point, of course, is that some tasks will require support under the responsibility of a team or department other than the one providing the principal engineer.

System test and evaluation verification tasks should be placed under the responsibility of a task principal engineer selected from what the author calls a program integration team (PIT), the manager of which reports to the program manager and may also be the chief engineer. The system test and evaluation function of this team would start small, of course, and grow toward the end of the engineering workload on the program. The work may have to be accomplished at an off-site location because of safety concerns. In military situations this work may even involve the delivery of live ordinance on targets requiring use of one or more special military test ranges. During one program the author worked on, it was necessary to include test flights across foreign countries and open ocean terminating in impact in target ranges in the western United States. In another case, flight was accomplished at high altitude over commercial passenger flight corridors to gain entry into Western test ranges for recovery. Engine flameout on one of these flights brought down a great deal of criticism on the program, but then very few people knew about it. These situations will require high-priced help to negotiate agreements.

Item acceptance principal engineers and item task principal engineers for those items should be selected from the responsible item team. Item task plans and procedures should be developed by task principal engineers, but the work itself will often be accomplished and reports prepared by manufacturing or quality engineers on the production floor using whatever degree of automation the enterprise manufacturing process employs. It is likely that the responsible development team will have been disbanded by the time production and acceptance have begun and certain that it will have been by the end of production. An alternative way to look at this situation is that cross-functional team leadership will have transitioned by this time from engineering leadership to manufacturing leadership.

It is important to note that the verification effort will require substantial data processing support. Ideally, a program would employ a database supporting both the requirements modeling and documentation as well as verification management in an integrated fashion.

1.3.1.8 Assembly of the Program Verification Process

There are many ways to assemble the verification process elements into a program plan as a function of the kind of program and methods of organizing the work employed. There appears to be nearly universal acceptance of the foursome of verification activities involving PMP verification, item qualification, item acceptance, and system test and evaluation, each married to a particular kind of specification. These four phases appear appropriate no matter how the verification work is assembled on a program, but there may be significant differences in how they are implemented. Chapter 5 begins the discussion of the verification process within a program.

1.3.2 Systems and Their Development

1.3.2.1 What Is a System?

We accomplish verification work on the design and manufacture of systems and their parts, but what is a system? A system is a collection of things that interact to achieve a predefined common purpose. There exist collections of organic and inorganic things in our universe that satisfy this definition, including the whole universe. In these cases, humans neither specify nor develop the system; we only attempt to describe the current reality and decide what collections of things in the natural environment to group together into specific systems of interest to us. This book and others by the author are primarily concerned with man-made systems that are created specifically to solve particular problems on or about planet Earth. A system is created in a solution space consisting of all of the possible ways a problem from a problem space defined in a specification could be solved. These man-made systems consist of physical and logical entities, which achieve predefined functionality exercising relationships, or interfaces, between them through which the richness of a system is achieved.

The need for a system is conceived by a person or organization with the intention or desire to solve a problem, the consequences of which can be observed or imagined in an environment of interest. If the problem is very difficult to solve, it may require a special organization to develop the solution. The needs for large systems are commonly conceived by what we call "users" of systems. These users commonly describe the problem in terms of their operational mission needs rather than in engineering terms. The organization that the user is a member of commonly cannot evolve a solution to the need statement from its own resources, so another agent is identified that is experienced in the acquisition of systems to solve user-observed problems. Acquisition agents in organizations such as the U.S. Department of Defense, FAA, and NASA study the availability of development enterprises that could produce a system to solve the need and let contracts with these enterprises to develop and deliver a system after a competitive proposal process through which competing contractors can bid to obtain the contract. These acquisition agents specify the need as precisely as they can and engage in a contract with the development enterprise to deliver a system that can be shown to possess the characteristics defined in the system specification. Thus is born a process called "verification" that seeks to establish the degree to which a developed system satisfies or complies with the requirements defined for that system. One view of verification is as a quality assurance provision.

1.3.2.2 Systems Development Overview

Systems come into being to satisfy specific needs and are commonly brought to life by enterprises that have come to be effective in the application of a collection of art, science, engineering, and good business sense to the solution of difficult problems brought to them by people and organizations in need of help with their problems. The enterprises discussed in this book are thought of as profit-making companies operating in the American economy, but the same principles would apply generally in a command economy.

1.3.2.2.1 What Is Systems Development?

Systems development is the art and science of creating man-made systems to satisfy predetermined needs. It is a problem-solving process where we bring to bear appropriate elements of mankind's knowledge base to create new knowledge specific to the problem and, as a result, define a solution to the problem. In this book, we refer to organizations that accomplish this work as system development enterprises, developers, or contractors that may, at any one time, be in the process of developing several systems, each through the exercise of an organizational structure called a *program*. When the problem is very complex, involving a knowledge base that is both wide and deep, and the customer wishes delivery within a constraining time period and cost limit, these enterprises find it necessary to employ many people in the development effort, leading to a need to manage the development effort well in terms of the capabilities of the evolving product, the number of billable hours accumulating, and the amount of schedule time consumed relative to a plan and contract.

The author is a devotee of the matrix organizational structure, in which a functional department structure is responsible for providing programs with the resources they need and each program organizational structure is responsible for managing the development work for one program using cross-functional teams assigned responsibility for specific product entities that will compose the system. Some enterprises are not large enough to make this structure serve their needs well and have to apply some combination of project and/or functional departments. There are also upper size limitations that are most often solved by an enterprise forming divisions, each focused on a particular product line functioning within a matrix structure.

In order for system development to be effectively accomplished, there are several casts of characters who must cooperate within an enterprise, and this is seldom achieved in practice at the level of intensity and duration needed, resulting in some loss of efficiency and profit. The enterprise must field a team of specialists for each program that can act as one against a common enemy of ignorance with the same intensity that a winning sports team does for their season. In order for this to occur in an enterprise employing a matrix organizational structure, functional department managers must be prepared to deploy to each program the best possible resources (personnel, practices, document templates, and tools), progressively improved over time based on program lessons learned and coordinated across the functional departments. In addition, the managers of programs must organize and manage these resources with great skill to achieve customer needs consistent with also meeting enterprise goals, including a profit. Program managers should also bring to bear on program and product problems experts in dealing with complexity, called system engineers, early in the program so that their skills can be most effectively applied while leverage remains most advantageous against future risk.

Other people must, of course, cooperate in the development effort from the many specialized disciplines or domains. These may include mechanical, electrical, fluid, and chemical engineers, but today they will also almost always include computer software domain specialists because software is an ever-increasingly important element of systems that solve very complex problems. While software is an intellectual entity, it can be developed for use in a computer to accomplish tremendously complex goals that will be very difficult to accomplish in hardware alone. It is true that a little more than half a century ago there was little software, and system engineers were hardware people because that is all there was. Many system engineers have remained hardware focused and some may even remain in denial that software exists. Anyone left in this situation today must expand their outlook to embrace software or perish as a system engineer. The reality is that everyone needed on a program provides a necessary service and the result will be less complete and satisfactory if they are omitted, but the author is a system engineer and this is a book about an important part of the system engineering workload on a program. So, perhaps the author can be forgiven what may appear to some readers as an unreasonable acceptance of the importance of the work of system engineers.

1.3.2.2.2 Three Steps on the Way to Great Systems

The systems approach applied by successful programs in solving complex problems and thus developing effective systems entails three fundamental steps: (1) define the problem in the form of requirements captured in specifications; (2) creatively synthesize the requirements into design solutions believed to satisfy those requirements, procure materials and equipment needed to produce the system, and manufacture the product; and (3) prove through test and other methods that the resultant product does in fact comply with the requirements, thereby solving the original problem. All of this work must be accomplished under the leadership and oversight of program and team management personnel within the context of a supporting overarching technical management environment respected and contributed to by system engineers, program management personnel, and functional management personnel. While some of the technical work accomplished in any of the three systems development steps on a program can be very challenging, none of it can be more difficult than the management of program personnel because managers have to work with the most complex systems that exist on planet Earth – human beings – interrelated by the worst interface on the planet – human communication.

An organization that develops products to solve complex problems is right to apply this organized three-step process to their work because it results in the best customer value, motivated by least cost and best quality in concert with needed performance. This happens because the systems approach encourages consideration of alternatives and the thoughtful selection of the best solution from multiple perspectives, including affordability, as well as the earliest possible identification of risks and concerns, followed by mitigation of those risks prior to them becoming serious program concerns. At the center of this whole process are the requirements that act to define the problems that must be solved by design engineers or design teams. It is only right that the developing enterprise should benefit from the good works derived from applying this process well and that will commonly be the case because the habits that result will attract customers to the enterprise.

This book focuses on step three in the systems approach, variously referred to as validation, verification, or V&V. We will see that there is no universal understanding of the meaning of these words, and the dictionary does not help very much. The author will use the term verification, as is obvious from the book title. But for the time being, let us say that these activities, by whatever name, produce evidence of the degree of compliance that the design possesses relative to its requirements.

Some people believe that the purpose of verification is to prove that the product complies with the requirements, but the author has observed that in some companies programs have produced false evidence that their product satisfied the requirements. Those who would verify must be people of the highest integrity. They must focus on truth rather than what management wants – even if it means you will be looking for another job. Any organization or manager who wants something other than the truth out of the verification process does not deserve to have people of integrity working for them and, if there is any justice in the world, they will not be employed for long in a position of trust and responsibility. What we want to understand is the degree to which the design truthfully complies with its requirements.

1.3.2.2.3 The Key Role of Systems Engineering

The solution of difficult problems flows from a apparently simple but actually difficult process that converts organized human thought, materials, and time into descriptions of organized sets of things that function together, to achieve a predetermined purpose that we call systems – the solution to the problem beginning the whole effort. This process is called the *system development process* in this book. It is an exercise in the practice of systems engineering supporting a systems development management effort and also supported by people from a large number of specialized disciplines. As mentioned, this process is actually simple in concept, but not easy to implement well because the related work must be accomplished by many people, none of whom can fully understand the whole problem in depth and breadth. In combination with the organization that implements the process and the people that populate it, the process becomes a complex undertaking. Our purpose in this book is to understand the verification part of this process, but to do so we must appreciate the whole to some extent.

There are some simple truths about systems engineering that, once appreciated, make everything else fall into place. The amount of knowledge that human beings have exposed is far in excess of the amount of knowledge that any one person can master and efficiently apply in solving problems. Developing systems is a problem-solving process that transforms a simple statement of a customer need into a clear, detailed description of a preferred solution to the problem implied in the need. Companies that are able to bring to bear more of the knowledge base related to a given complex problem will have the best chance of success in solving these problems. Over time, this ability can be translated into the basis for a successful business venture in the development of systems that solve complex problems. Note we have been very careful to avoid the complex system construct. Many systems are very complex, but that should not be our purpose. We seek to solve complex problems with the simplest possible systems that will satisfy the need.

Most of the simple problems faced by humanity have already been solved somewhere in the world. Many of the problems we now conceive represent fairly complex new problems or extensions or combinations of previous solutions applied to a grander, more general problem space. Complex problems commonly require an appeal to multiple technologies and a broader knowledge base than any one person can master. In order to accumulate the knowledge breadth necessary to deal with a complex problem, we must team more than one person together, each a relatively narrow specialist who deeply understands their specialty, and organize their work environment to ensure that synergism occurs between the several kinds of specialists. *Synergism*, a key systems term, is the interaction of elements that when combined produce a total effect greater than the sum of the individual elements. Synergism is necessary to weld together the individual specialized knowledge spaces into an aggregate knowledge base that the team needs to solve the problem. At the same time, a team may fall into a lucky condition of serendipity resulting from the application of creative genius or dumb luck that should not be cast aside by systems purists.

We are clear on what the specialized engineers do within process systems in the development of product systems. They ply their specialized trade in the definition of product requirements and development of design features related to their specialty. While success in this work is a necessary condition for the success of the overall project, it is also true that these specialized engineers cannot solely among themselves accomplish the development of a complete product system, because of their specialization, without the cooperation of generalists who are more interested in the whole than the parts. We will call these people *system engineers*.

Specialists have a great deal in common with Sherlock Holmes, who told Watson in the first of those unforgettable stories that he had to guard against admitting extraneous information into his mind for it would crowd out information important to his profession. If an enterprise were to choose to expand and generalize the knowledge base of their specialized engineers to eliminate the need for any overlying coordinating function, they would, in general, become less competitive because the specialists will not, over time, be as sharp in their special fields as their more finely focused counterparts in competing firms. You cannot form a successful company composed entirely of system engineers. You must have some people who know a great deal about each domain. One of the important roles of a system engineer is to integrate and optimize across the work of the many specialists. So system engineers must work at the boundaries and be good communicators.

Some time during 1968 in the officer's club at Bien Hoa Air Force Base, South Vietnam, the author once debated the relative merits of specialization and generalism with an Air Force captain for several hours. The captain was the unit maintenance officer who claimed to be specialized in management of personnel. At the time the author was a field engineer representing Ryan Aeronautical Company serving with a Strategic Air Command detachment that was operating U-2 aircraft and a series of unmanned photo reconnaissance aircraft developed by Ryan. These unmanned aircraft were carried aloft out of Bien Hoa on DC-130 launch aircraft and launched into North Vietnam and China. His job, 10,000 miles from the plant and often alone, was to support the customer for all systems in all active models and related launch aircraft equipment for maintenance and operations purposes, so by necessity, he drifted into an appreciation of generalism. The captain ended the conversation that evening with the comment, "Jeff, I'm not a tech rep so I have to get up in the morning. So, let me see if I've got this straight. I think you are specializing in generalism." As deflating as this comment was at the time, the author has since grudgingly concluded that the captain was absolutely correct. We are all specialists with a different balance point between knowledge breadth and depth, as illustrated in Figure 3.

We have a choice of knowing a great deal about one or a few things, or a little about many things. This is the basis for the oft-quoted statement about

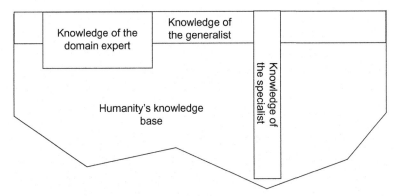

Figure 3 We are all specialists.

the ignorance of system engineers relative to specialists: "System engineers are a mile wide and an inch deep." Yes, it is true, and it is necessary. The alternatives, qualitatively clear from Figure 3, are that individuals may apply their knowledge capacity in either direction, but not both in a comprehensive way; we humans are area limited on this figure and in life. In the development of systems, we need both the narrow specialists to develop ideas appropriate to the solution of specific problems and generalists skilled in fitting all of the specialized views of a system into a whole. In very complex problems and systems, this hierarchy of knowledge granularity may require three or four distinct levels.

Over time, we have devised a very organized process, the systems approach, for solution of complex problems, which orchestrates a progressive refinement of the solution derived by a group of specialists, none of whom are individually capable of a complete solution. A complex problem is broken into a set of smaller problems, each of which can be efficiently attacked by a small team of specialists. To ensure that all of the small-problem solutions will combine synergistically to solve the original complex problem, we first predefine essential solution characteristics, called requirements, which are traceable to the required characteristics for the larger problem, all of which are derived from the customer need, the ultimate requirement.

The resultant logical continuity gives us confidence that the small-problem solutions will fit together to solve the large problem. Requirements validation is applied during concept development and early design work to ensure that specific requirements are necessary, optimally valued, and that it is possible to satisfy them with available knowledge and resources. Verification work is applied subsequent to the design work on test articles and early production items to produce evidence that the design solutions do, in fact, comply with the requirements.

The rationale for decomposing large problems into many related smaller ones is based on two phenomena. First, as noted above, a very large problem will commonly appeal to many different technologies. Complex problem decomposition results in smaller problems, each one of which appeals to a smaller number of different technologies than necessary for the large problem. Therefore, fewer different specialists are needed, and the coordination problem is reduced from what it would be for direct in-depth solution of the larger problem. Second, breaking large problems into smaller ones encourages team size and granularity consistent with our span of control capabilities. The systems approach is, therefore, driven by perfectly normal individual human problems of knowledge and span of control limitations. It is a method for organizing work of a complicated nature for compatibility with human capabilities and limitations.

It is important to understand that, whenever we apply decomposition to a problem space, we must also apply integration and optimization to the parts into which we decompose the whole, in order to encourage the smaller elements to interact synergistically when completed. Note that we partition both knowledge and the solution space in developing solutions to problems. Integration is necessary across both of these boundary conditions. System engineers live at these boundaries encouraging effective communication between the specialists.

To master all knowledge, we have found it necessary to specialize. Because we specialize, we need an organized approach to promote cooperative action on problems so complex that no one person can solve them. System engineering work is very difficult to do well because it does require cooperative work on the part of many people, whereas specialists generally have the luxury of focusing on their own specialized knowledge base and vocabulary. System engineering work entails an outgoing interest direction from the practitioner, while specialized work tends toward an inward-looking focus. System engineering promises to become even more difficult to perform well in the future because our individual knowledge limitation is fairly fixed and our knowledge base continues to expand. Our collective solution to that continuing problem can be observed in Figure 3 in the form of emerging domain system engineers for various fields like telecommunications, software, and control systems. This growing problem is made more serious because more refined specialization is relatively easy to implement, but the compensating improvements in our ability to integrate and optimize at higher levels is more difficult to put in place. The former is oriented toward the individual. The latter requires groups of people and is thus more difficult at least in part because of the difficulty of communication across domain and product entity boundaries. As a result, a system engineer must also be effective in communicating ideas across knowledge boundaries.

It is this difficulty, based on many people cooperating in a common responsibility, that has caused some development organizations to evolve a very rigorous, serial, and military-like or dictatorial environment within which the work of humans is forced into specific pigeonholes clearly defined in terms of cost and schedule allocations, and the humans themselves devalued in the opinion of some. The original intent for the systems approach was to encourage cooperative work among many specialists concurrently, but in practice the process has not been applied well in many organizations focused on functional departments and a serial work pattern. This book encourages what is often called the concurrent development process, performed through cross-functional teams organized about the program product entity structure on programs with functional department support in terms of supplying programs with trained personnel, good written practices, and good tools, all incrementally improved in a coordinated way over time based on lessons learned from programs and the results of independent research and study.

It is understood that people, once assigned to the program from their functional departments, will be organized by the program into product-oriented teams and managed on a day-to-day basis by the program and not functional management. This is necessary in order to avoid the single most difficult problem in the development of systems: cross–organizational interface responsibility voids. We organize by product to cause perfect alignment between those responsible for the work and the program planning and resources they will be supported by, but also in the process we see that a one-to-one correspondence develops between the teams and the product entities for which they are responsible. However, responsibilities for interfaces between product entities are shared between team pairs and the human communication patterns that must take place to resolve these interfaces pose special problems, because the responsible teams tend to look inwardly at their end of the interface. System engineers provide integration support between these team pairs to encourage compatible interface terminals.

The work described in this book requires the cooperation of many specialists involved in different technology disciplines from many company functional departments. This work requires the capture of a tremendous amount of information that must be accessible to and reviewed by many people. The amount of information involved and the broad need for access suggests the use of distributed databases within a client-server structure, and that conclusion will be reinforced throughout the book.

1.3.2.3 System Modeling Alternatives

Chapter 3 briefly covers four universal architecture description frameworks (UADF), any one of which can be used by an enterprise on all of its programs to determine an appropriate system architecture, its product and interface composition, and requirements to populate all program specifications no matter how the design will be implemented in terms of hardware, software, and people doing things. The full story on these modeling methods is covered in a companion volume to this book titled *System Requirements Analysis* by the same author and publisher. These four models are: (1) functional, (2) a combination of modern structured analysis and process for system architecture and requirements engineering (MSA-PSARE), (3) a combination of unified modeling language and system modeling language (UML-SysML), and (4) a unified process for DoDAF-MoDAF (UPDM). In the latter three cases, models for specialty engineering and environmental requirements have been added to some already fairly complicated models. UPDM includes 52 modeling artifacts but is still incomplete as a requirements model. Ideally an enterprise would apply the same model to address all three system components with which every enterprise must deal: (1) the product system, (2) the program system that will develop a particular product system, and (3) the whole enterprise containing all currently operating programs and a functional department structure.

1.3.3 Characteristics of the Enterprise

The enterprise and its programs and how they are organized have some influence on how well the work of verification can be done. The author, like all other people, has a preference that the reader should be made aware of.

1.3.3.1 Development Enterprise Organizational Structure

A common way for large enterprises to organize to accomplish work on multiple programs is to apply the matrix management structure. One axis of the matrix consists of a functional department structure, where the managers of each department are responsible for maintaining and improving the knowledge base of the enterprise and supplying programs with good people who know what they are doing, good practices, and good tools, all coordinated across the enterprise. The enterprise should have a generic or common process developed and maintained by an enterprise integration team (EIT) that integrates and coordinates functional department contributions in accordance with clear responsibilities derived from allocation of enterprise generic functionality to the functional departments.

A system development enterprise accomplishes its work through programs, the other axis of the matrix. The assumption in this book is that the enterprise must operate multiple programs at any one time, requiring the common functional organization structure to provide resources to all of these active programs. Figure 4 illustrates the author's preferred expression of the matrix organizational structure. The diagram is expressed at what is commonly called a division level, where the enterprise may have multiple divisions, each focused on a particular product line. Each of these divisions is

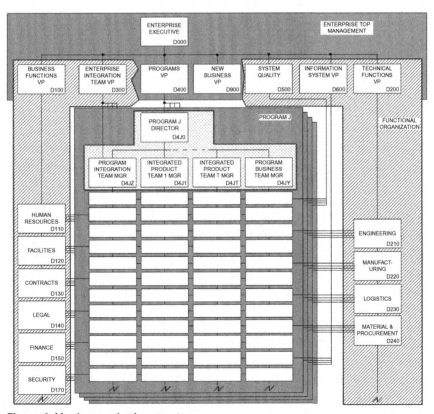

Figure 4 Matrix organization structure.

essentially a process-oriented system, as are each of its programs. The system entities in these cases are humans and organizations of humans and the relationships between them are the worst interface on planet Earth – human communication.

The functional organization is split into technical and business entities, each of which provides resources (people, practices, and tools) to multiple program team structures. The author split the functional structure into these two components to graphically balance the picture and there is no real need to recognize this distinction. Programs come and go, and the resources flow from one to another through the functional department homerooms. An EIT acts as the enterprise integration and optimization agent, the system engineer for the enterprise. The personnel assigned to the programs are managed through cross-functional product entity structure oriented teams on the programs.

The programs begin with a program integration team (PIT) and program business team (PBT). As the PIT identifies an appropriate product entity structure, using methods described in the author's book *System Requirement Analysis* for which an overview is offered in Chapter 2, it must identify a team structure oriented toward that product entity structure and staff these teams from the functional departments as a function of the kinds of work that will be required in each team.

The process flow diagrams used in this book were crafted employing a functional universal architecture description framework (UADF), but an enterprise could employ any one of four UADF modeling approaches covered in Chapter 2. Ideally, an enterprise would select one of these and apply it on all programs. Each integrated product team (IPT) leader is given a specification (crafted by the PIT or higher tier IPT), program planning data, budget, and schedule and then set to work and managed in accordance with that planning. The PIT (the program system team by whatever name) acts as the program integration and optimization agent supporting the program manager. Each IPT must act as the integration and optimization agent for any activity subordinate to the team level item, especially where subteams are assigned.

1.3.3.2 The Business of Programs

The techniques described in this book can be applied at any level of indenture in a system, from the system level down through the component level. So, when the author speaks of "systems," the reader can understand it to mean whatever level that corresponds to the reader's scope of interest.

The content also applies to hardware and software alike, though the reader should be aware that the author's principal experience has been in hardware beginning before software existed. Some of the book content has been prepared specifically for hardware or software applications, but most of the content applies to both.

The content also applies whether the product is a new weapons system for the Department of Defense, a new automobile model using electric power, or a new farm tractor. This approach owes its emergence to the many military and space systems developed in the United States and other countries roughly during the period 1950 through 1990, but it is appropriate for any development effort intended to solve a very complex problem. This process may be viewed as a part of a risk management activity to encourage the identification of problems while there is still a lot of time to solve them or to mitigate their most adverse effects. If you choose to accept a lot of risk, you may not feel obligated to apply these techniques. But, you should steel yourself to deal with a lot of very difficult problems under stressful conditions resulting in unplanned cost, late delivery, and reduced product effectiveness. Risks are not just idle curiosities discussed in academic treatises. Many of them will be fulfilled during the life cycle of a system, even one that was well managed during development.

There are ways to economize in the application of the systems development process discussed in this book. One need not necessarily apply the full toolbox of techniques discussed, or to the level of intensity discussed, to be at least reasonably effective in reducing program risk. This systems approach entails the following principal steps employing a functional modeling approach during requirements analysis (see the author's *System Requirements Analysis*, also from Elsevier, for alternative modeling approaches):

a. Understand the customer's need – that is, the ultimate system function, for it tells what the customer wants achieved. It defines the problem that must be solved at a very high level and not very precisely. This may be stated in a sentence, a paragraph, or a very few pages.

b. Expand the need into a critical mass of information necessary to trigger a more detailed analysis of a system that will satisfy that need. This may entail customer surveys, focus groups, question-and-answer sessions, or mission analysis in combination with a high-level functional analysis that exposes needed functionality, system design entailing allocation of the top-level functionality to things placed in a system physical product entity structure, and definition of requirements for the complete system (commonly captured in a system specification). Through this process, one will determine to what extent and for what elements solutions shall be attempted through hardware and software.

c. Further decompose the need, which represents a complex problem, within the context of an evolving system concept, into a series of related smaller problems, each of which can be described in terms of a set of requirements that must be satisfied by solutions to the smaller problems. Translate the identified functions into performance requirements and allocate each to a product entity. Complete design constraints modeling work and identify related requirements (interfaces, environmental, and specialty engineering) for the product entities.

d. Prior to the start of detailed design work it is sometimes necessary or desirable to improve team member confidence in their understanding of the requirements or to prove that it is physically possible to produce a design that is compliant. This is commonly accomplished through special tests, analyses, and simulations, the results of which indicate the strength of the currently available related knowledge and technology base and the current state of readiness to accomplish design work using available knowledge that will lead to success. This work is referred to as validation in this book and is covered in the author's book titled *System Requirements Analysis*, also published by Elsevier Academic Press.

e. Apply the creative genius of design engineers and the market knowledge of procurement experts within the context of a supporting cast of specialized engineers and analysts to develop alternative solutions to the requirements for lower-level problems. Selections are made from the alternatives studied based on available knowledge to identify the preferred design concept that will be compliant with the requirements. Product designs are transformed into procurement and manufacturing planning and actions, resulting in the availability of a physical product and related software.

f. Integration, testing, and analysis activities are applied to designs, special test articles, preproduction articles, and initial production articles that prove to what extent the designs actually do satisfy the requirements. It is this verification process that this book focuses on.

This whole process is held together, especially during the initial theoretical stage, by requirements. These requirements are statements that define the essential characteristics of a solution to an engineering problem prior to the development of the solution to the problem. This book is about requirements and determining the extent to which they have been satisfied in designs. Requirements are identified as a prerequisite to accomplishing design work because of the nature of complex problems and the need for many people to work together to develop the systems that can solve these complex problems. Thoughtfully identifying and capturing all of the requirements for a given product entity in a specification is a means of communicating across the team members what features and capabilities the design must possess.

System engineers should never lose consciousness of the fundamental responsibility they have to their fellow, more specialized, engineers relative to communications. The proper residence of the system engineer on a program is at the knowledge and product boundary conditions, which happens to most often coordinate with difficult conversations that the specialists must have and product interfaces they must mutually understand. All too often the specialists will withdraw from these boundaries and focus internally on the problems they understand more fully. System engineers have to insist on the parties engaging in communication no matter how painful that can be sometimes.

1.3.3.3 Program Structures

Customers for very large and complex systems like the DoD and NASA have, over a period of many years, developed very organized methods for acquisition of their systems. Commonly, these methods entail a multiphased process that partitions complex problem solving into a series of steps or program phases, each ending in a major review where it is determined to what extent the planned progress has been achieved before committing to expenditure of additional funding to carry on into the next phase, and where each phase involves increasing funding demands and potential risk of failure due to past undetected bad choices. The purpose of the phasing is to control program risk by cutting the complete job into manageable components with intermediate decision points that allow redirection at critical points in the development process. We see this same practice on a smaller scale with a $50,000 home remodeling effort, when the contractor agrees to a contract that pays him 40% up front, 30% based on some significant achievement, and the remaining 30% at completion.

Since large agencies like DoD and NASA that are customers for large systems are government bodies, they must function within a growing volume of laws, rules, and regulations intended to preserve a level playing field for businesses seeking to satisfy their needs, maximize the benefits to the country from procurements, and discourage failed procurements. The use of public funds properly carries with it burdens with which commercial enterprises need not contend. Therefore, commercial enterprises should properly avoid some of the extensive oversight and management methods applied to government programs, but they should think carefully before they throw the whole process out the window, including the technical and management support processes useful in both military and commercial product development.

Figure 5 offers a generic phasing diagram suitable for companies with a varied customer base with overlays for DoD and NASA procurements as they were a few years ago. Each phase is planned to accomplish specific goals, to answer specific questions, and control future risks. Each phase is terminated by a major review during which the developer and the customer reach an agreement on whether phase goals have been satisfied. Both DoD and NASA have evolved their own life-cycle process diagrams since this diagram was first created, but the pattern is unchanged, suggesting that an enterprise can evolve a common life-cycle model serving all of its customers, based on a common process that its employees can focus their work upon and can continue to improve their performance individually and collectively on programs.

Generally, the information developed in a phase is approved as a package and used to define a baseline serving as a platform for the subsequent phase work. In this way, controlled progress is made toward solving the original complex problem represented by the customer need. Figure 5 applies what is called a waterfall model and later we will consider alternative ways of viewing the program implementation.

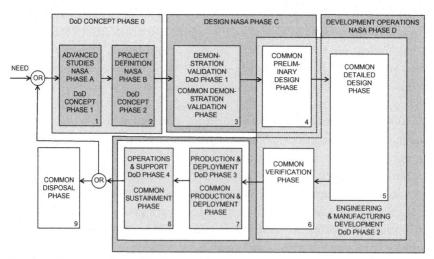

Figure 5 Generic systems development process.

The development of commercial products has commonly been accomplished by companies in what appears to be a very different environment from that described in this book. There appears to be, among engineers and managers in commercial enterprises, a great dislike of the traditional DoD-inspired systems approach to complex problem solving. The author is a student of these attitudes but has not mastered what he believes to be the real basis for it. In talking to people in the commercial development world about their product development process, we find one of two possibilities to be in effect: (1) they cannot explain their process, suggesting that it is an ad hoc process that just happens, fueled by the personalities at work and the thoughts of the moment; or (2) they explain a process that most anyone experienced in the DoD approach could identify, with possibly some semantic differences.

The degree to which the process is predefined aside, the commercial world commonly does not have the benefit of a customer who speaks to them clearly about their needs. Some experienced in early program work on military contracts would argue that this does not often happen with large government customers either, especially as it relates to the clarity of the message. But it is true that DoD and NASA developments proceed from a customer position on the need with generally one voice, though sometimes users and acquisition agents differ on the operational need achievable with available funding. In commercial developments, the customer is seldom a homogeneous whole with a single spokesperson. The developer must make a real effort to uncover who the customer is and what will sell to that customer base. They must play the part of the customer as well as the producer. They must have a built-in voice of the customer that accurately reflects real customer interests.

The author maintains that the program spark may be different between the DoD and commercial worlds, but that what evolves thereafter need not be so very different. The fundamentals of the DoD-inspired approach are valid no matter the situation so long as we are interested in solving complex problems that require the knowledge of many people in specialized disciplines. The most efficient way to solve such problems that we currently understand is to decompose them into a related series of smaller problems as already discussed, assign the right collection of experienced people to those small problems, and apply system integration and optimization across the boundary conditions. So, it is problem complexity and not the nature of the customer that defines the need for an effective system engineering process.

We could also dig deeper and inquire just how different the beginning is between commercial and government programs. In a commercial enterprise, particularly in the consumer products area, a good deal of effort must be made to try to understand potential market opportunities in terms of what will sell and who might buy their products. This is done through product questionnaires received with the product, focus groups, market studies, analysis of statistical data from government agencies and polling firms, historical trends, and lucky guesses.

If you scratch the surface of a successful DoD contractor, you will find activities that you might expect to be going on in a commercial firm to understand the customer. You will find the marketing department and its supporting engineering predesign group working interactively with one or more customer organizations, either cooperatively or with some separation, to understand the customer's needs, offer suggestions for solving those they perceive, and to remain abreast of customer interests. This is the basis of the often-repeated saying, "If you first become aware of a program in the *Commerce Business Daily (CBD)*, it is too late to bid." Federal government agencies are required to announce any procurement over a certain dollar value in the *CBD*, which comes out every workday and is now available from a Web site rather than in paper format. The rationale behind the saying is that the future winner of the program noted in the *CBD* probably helped the customer understand that the need existed through months or years of close cooperative work and related independent research and development (IRAD) work to control technical risks and demonstrate mastery of needed technologies.

So, the fundamental difference between DoD and commercial development may be as simple as the degree of focus and leadership within the two customer bases. In the commercial world, especially for consumer products and less so for large commercial projects like large buildings, commuter rail line construction, and oil exploration, the voice of the customer is scattered and uncoordinated. In DoD procurements, the voice of the customer is focused at a point in the government program manager. Yes, it is true that there are many different expressions of the DoD customer voice in the form of the procurement agency, government laboratories, and multiple users, but the customer voice is focused in the request for proposal or contract under the leadership of the program manager. This is not commonly so in commercial industry.

1.3.3.4 Toward a Standard Process

Enterprises evolve over time to serve a particular customer base, offering a relatively narrow product line. In most enterprises, even those with multiple kinds of customers (government and commercial, for example) a single generic process can be effective. This is a valid goal because the employees will progress toward excellent performance faster through repetition of a sound common process across several programs than when they must apply different techniques on different programs. It is also difficult to engage in continuous improvement when you are maintaining more than one fundamental process. A single process encourages development and improvement of personnel training programs, documented practices, and tools keyed to the training and practices. It is hard to imagine why one enterprise would want to maintain multiple processes, but some do, reinventing themselves for each new proposal based on bad feelings about previous program performance.

This is not a simple situation, as suggested by Figure 6. Our enterprise should be guided by an enterprise vision statement (by whatever name) that succinctly states our fundamental purpose or goal. This is the ultimate functionality of our enterprise and can be decomposed using the same modeling techniques we would apply in the development of product systems. These methods are briefly discussed in Chapter 3 fitted into a detailed generic process diagram. That generic diagram can then be coordinated within the enterprise, managed through a matrix structure with the functional departments

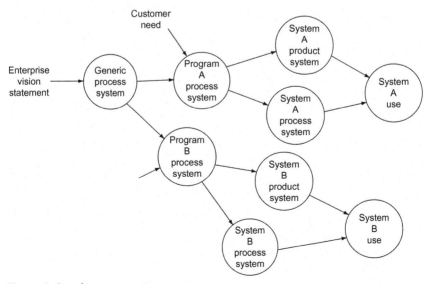

Figure 6 Grand systems pattern.

(illustrated in Figure 4 on left and right sides) that must supply program resources. The departments develop the generic practices, acquire the tools, and train the employees in both while continuing to optimize the enterprise's knowledge access related to their product line and customer base.

Each program that comes into being should reflect the generic process definition. During implementation, each program should continue to improve the details about the program process while developing the design for the product and its implementation process. This book is one of four books written by the author in support of a system engineering certificate program that his company has offered for 20 years, concerned with optimizing the aggregate of these elements, a single generic process and multiple program processes all optimized at the enterprise level and each respecting an integrated and optimized product development process. Each entity in this structure requires an integration and optimization agent to encourage success for that entity. Each program should have a program integration team, as shown in Figure 4, and it should be the task of this team to encourage an optimum integrated product and process.

The enterprise should have an enterprise integration team, as also shown in Figure 4, to encourage that all programs be jointly optimized about enterprise objectives. The aggregate of the enterprise and all of its programs is referred to as a *grand system* for which the enterprise integration and optimization team is intended to be the systems engineering agent.

Figure 7 provides one view of a generic process diagram emphasizing the product and process development activity. This process follows the traditional sequential, or waterfall, model, at a very high level, involving requirements before design and verification of design compliance with those requirements. One must recognize that on any given program, at any given time, some items may have completed the requirements analysis process and have moved to the design process while other requirements work continues for other items. A pyramidal structure is appropriate here, suggesting that the concept development work starts at the system level, the peak of the pyramid, moves down to the next level under two or more teams, and continues to grow in terms of the number of teams simultaneously involved but at different points in the development pattern. Once the requirements are approved for any of these activities, the responsible teams can begin an upward movement focused on design, integration, and optimization.

Figure 7 provides a phasing diagram within which a standard process functions to achieve program phase goals. Figure 8 illustrates how these two views, generic process and program phasing, correlate. In each phase

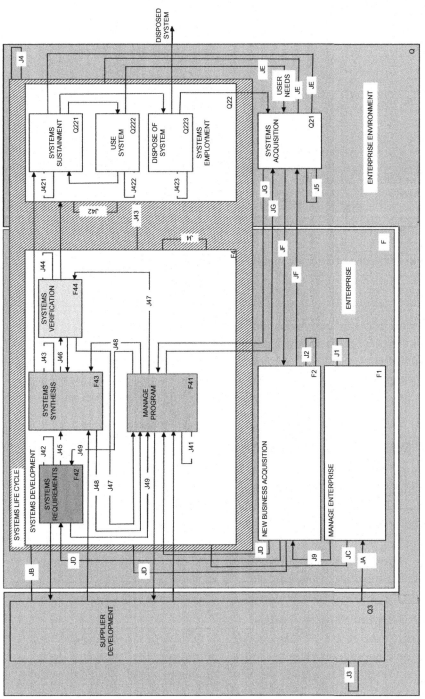

Figure 7 A common enterprise life-cycle model.

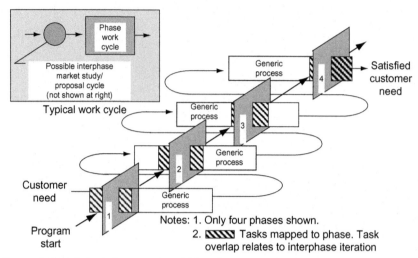

Figure 8 Correlation of a process model and program phasing.

we accomplish a degree of development work by applying our standard process for a particular system depth and span. The next phase may require some iteration of previous work based on new knowledge suggesting a different approach, providing for better system optimization. In each phase we repeat some of our generic process steps mixed with some new processes not previously accomplished at a more detailed system level.

1.3.4 Development Environments

This book follows the traditional waterfall model in general employing a functional flow diagram process depiction and encouraging adherence to a plan and schedule, but there are other environments within which a program may find success. We will look briefly at several variations that the reader may wish to consider: four program phasing models, two process discipline variations, three variations in the number of systems delivered, and four system modeling variations.

1.3.4.1 Program Phasing Models

The single process that is being encouraged in this book could employ any one of several program phasing models. Up through the 1960s and 1970s, there was only one accepted system development model. That was what we now call the waterfall model and that is what the discussion up to this point has reflected. Since then, the software world has evolved alternative models that have proven useful in the development of software under some circumstances. These include the spiral and V models. The author has added a variation on the V model, called the N model. In all of these cases, one should note that at the micro level the same cycle of activities is taking place: requirements analysis, synthesis, and verification within the context of a sound management infrastructure.

This cycle is at the heart of the systems approach. In some models discussed in the following sections, we intend that we will complete each of these steps in one pass. In others, the understanding is that we will have more than one opportunity to complete any one of the steps. But in all cases, we will pass through each step in the same order. In applying these models, we can choose to apply the same model throughout a single system development activity, or we can choose to apply one model at the system level and one or more others at the lower tier in the same development activity. So, for a given development activity, the model may be a composite, created from the models discussed here, that is very difficult to illustrate in two dimensions, but imaginable based on the pictures we will use.

1.3.4.1.1 The Waterfall Development Model

The most commonly applied development model is called the waterfall model because of the analogy of water falling from one step to the next, as illustrated in Figure 9. In its pure configuration, each step ends abruptly, coinciding with the beginning of the next step. In reality, these steps overlay because, when the requirements are approved for one item, its design work may begin, while other requirements analysis work continues for other items. This overlay condition exists throughout the development span. Some people prefer to invert the sequence such that the water falls uphill, but the effect is the same. Figure 9 simplifies the process to some extent in that it is necessary to produce some product to run verification testing upon, as well as subject every production article to some form of acceptance test.

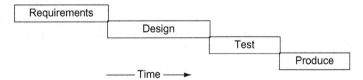

Figure 9 The waterfall development model.

1.3.4.1.2 The Spiral Development Model

The spiral model, shown in Figure 10, illustrates the growth of information about the system between the beginning, expressed in the customer need statement, and delivery by virtue of the expanding diameter of the spiral in time. For any given item being developed, the development team first defines requirements to the extent they are capable, does some design work based on those requirements, and builds a prototype. Testing of that prototype reveals new information about appropriate requirements and design features not previously exposed. This process may be repeated several times until the team is satisfied that

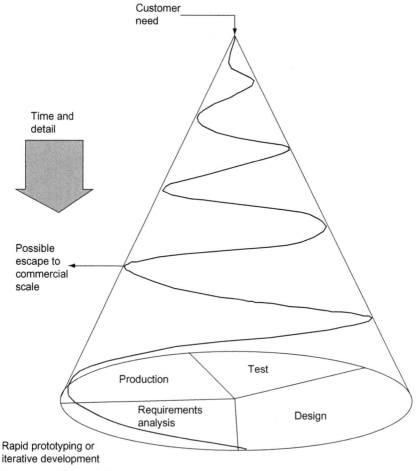

Figure 10 The spiral development model.

the product design is satisfactory. There are several development management models that have evolved from software development experience involving multiple cycles and distributed management responsibility that the author is frankly not a fan of.

This model is attacked by some as leading to point designs without benefit of an adequate requirements analysis activity and clear definition of the problem. That may be a valid criticism, but there are situations where we have difficulty faithfully following the structured approach suggested in this chapter using the waterfall approach. One of these situations involves the development of computer software that must be very interactive with the actions of a human operator. It is very difficult to characterize an appropriate set of requirements for such a system because of the subjective nature of the human responses to a particular stimulus. The development can be made to flow much more rapidly if the team has an opportunity to physically experience one or more alternative concepts and feed back the results into the next development iteration.

This model is also useful in describing software development through a sequence of versions. The team may conclude at some point in the development that the product is competitive relative to the products in development by rival firms, and, even though it may be imperfect, it can be released to beat the competition to market. Subsequent to a version 3.2.3 release, work continues to evolve improvements and fixes for known bugs that the customer may or may not be expected to encounter with the present release. Between product releases, customer service provides phone help to persons unfamiliar with the residual bugs and how to prevent them from becoming a bother. A product may remain in this development process for years, as in the case of the word processor used in the creation of this book. The product support answer is always the same no matter how many bugs have been dealt with at shipment time and how many more are to come: "Just wait till version 3.2.4. It is going to knock your socks off."

At the time this book was published, DoD had selected the spiral model as its preferred development model, to match the realities that DoD is forced to consider cost as an independent variable and on a particular development to make as much progress toward its ultimate needs as possible, consistent with available funding from Congress. One finds interest in DoD program offices in the development of a good chassis for the M1-A1 tank that can continue to be evolved over time in the form of software and electronics upgrades.

1.3.4.1.3 The V Development Model

The development process can be described in terms of a development downstroke and an upstroke. On the downstroke, we decompose the need into a clear definition of the problem in terms of what items the system shall consist of and the needed characteristics of those items. At the bottom of that stroke, we design the product in accordance with the requirements defined on the downstroke. Sufficient product articles are manufactured to test in accordance with test requirements defined during the downstroke. The test results qualify the product for use in the intended application. Figure 11 illustrates the V model, which is especially effective in exposing the close relationship between the requirements analysis process and the verification activity to prove that the design satisfies the requirements.

This is a very simple view of the development process. In actuality the system expands on the second level into two or more end items or configuration items, each of which expands into multiple subsystems and components. So, the V actually involves a shred-out on the downstroke and an integrating stream on the upstroke that is not clear from a simplistic

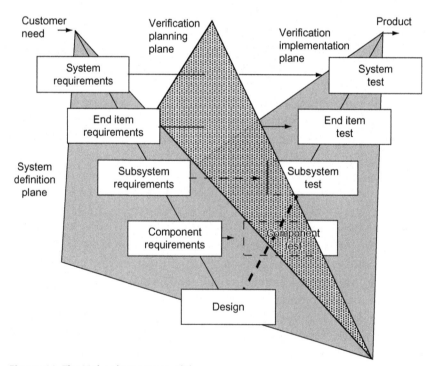

Figure 11 The V development model.

two-dimensional diagram. For a more expansive display of this view, see a paper by Kevin Forsberg and Harold Mooz titled "The Relationship of Systems Engineering to the Project Cycle" in *A Commitment to Success*, the proceedings from a joint symposium sponsored by the National Council on Systems Engineering and the American Society for Engineering Management in Chattanooga, Tennessee, in October 1991.

The completion of the functional and physical modeling plane work clearly defines the problem that must be solved in the form of a collection of specifications. We assign responsibility for development of the products represented by the blocks of the product entity diagram to teams at some level of granularity to accomplish the integrated design work for those items. These are the small problems we defined as a result of decomposing the large problem represented by the need. As the design team product designs become available, we manufacture test articles and subject them to planned tests and analyses to determine if they conform to the requirements previously defined. The final result is a product that satisfies the original need.

The author prefers to split the downstroke of the V model into two components to emphasize the fundamental difference between the functional and physical models useful in characterizing systems under development. This results in a forked V or an N model, illustrated in Figure 12. The functional

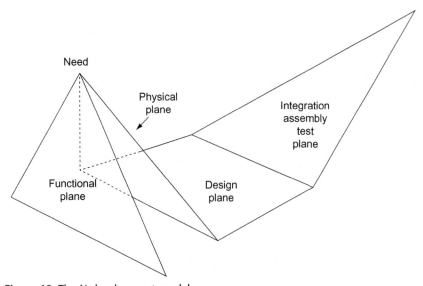

Figure 12 The N development model.

plane corresponds to the functional analysis process through which we gain an insight into what the system must do. For each function we are obligated to derive one or more performance requirements that are allocated to product entities that appear on the physical plane. The performance requirements are defined on the functional plane, but the remaining requirements, the design constraints, must be identified for each entity and must be derived from a trio of models on the physical plane. The functional modeling terminates when we have completed our product entity definition to the degree that we understand the problem in terms of all of the items at the lowest tier of each branch that will either yield to purchase from vendors or yield to design actions by a team of people in our own enterprise.

1.3.4.2 Process Discipline Variations

One can apply any of the systems development models rigorously following a phased sequence of activities or allow the creative juices of team members to express themselves in a more open environment. Some would say the latter is more fun, but it is also more open to potential disaster.

1.3.4.2.1 Ridged Process Compliance

This book favors a fairly ridged conformance to the traditional formal development process and describes the implementation of verification in that context. In this case the requirements for a particular system entity are fully developed prior to a team beginning the detailed design process and that design matures through an effort of several team members, each with different specialized knowledge discussing alternative design concepts and reaching a team decision about the relative merits of each, selecting a preferred concept that is then expressed in engineering drawings or code.

1.3.4.2.2 Rapid Prototyping and Rapid Application Development

The term *rapid prototyping* is commonly applied to the development of physical entities employing computer aided design (CAD) techniques followed by building a prototype and evaluating against some kind of criteria. This process may be applied in a spiral fashion until an adequate solution is perceived or available money runs out. With the availability of supporting equipment and software, the CAD output can be coupled to a computer application that translates the three-dimensional design into layers, which are then used to drive a three-dimensional printer that actually generates an operating physical reality reflecting the design. This entity in some cases can actually be adequate for sale and use by the public, but if the entity will have to endure substantial stresses the structure of it may not withstand those stresses for a very long time. But, in any case, the article thus produced can be very helpful to the team working on the development of a final product to validate the requirements showing that a design is possible, provided that the team actually uses the requirements as a basis for the design captured in CAD. The author believes that a lot of people and organizations employing this technique do so with little interest in an effective requirements analysis activity preceding the design effort.

This approach applied in the software world is often referred to as *agile development*. Team members work creatively to blend their skills and knowledge together, building on good ideas that some members express. It is claimed that in the development of solutions to very complicated problems it is not uncommon that the development team often gains insight into emerging behavior as the consequences of the intended solution becomes more clear, while an understanding of the difficulty of the problem continues to mature. Following a rigorous development model in these cases it is said that the program will have to spend a great deal of time unwinding the program from decisions reached before the true nature of the problem space had been clearly identified.

Some of the risk can be reduced through an effective risk management program and use of trade studies to evaluate the relative merits of several alternative choices. Some do criticize the rapid prototyping approach as appealing to those who don't see the benefit of an effective requirements analysis activity.

Item qualification in these cases can be applied to the prototypes created but may not provide useful information about requirements compliance from a structural perspective, where physical objects are generated. In any case, as a program employing these techniques translates prototypes into real products, those products can be subjected to all three phases of verification and a program can come to a conclusion with some money and time shaved off from that required for a more ridged conformance to the formal process, but unforeseen risks are often realized.

1.3.4.3 Number of Systems Delivered Variations

The principal matters that dictate verification process differences are: (1) the number of articles to be produced, (2) the cost of the articles, and (3) risks anticipated in system use. Subordinate paragraphs explore three specific situations and the circumstances in which they would be applicable. This book primarily discusses verification in context with the high rate production program pattern but the principal steps apply in the other two cases as well.

1.3.4.3.1 High-Rate Production Program

Where many product articles are going to be produced, it makes sense to call for a dedicated set of engineering units to be manufactured, recognizing that these items will be consumed in item qualification, followed by production of the articles to be delivered to the customer. If the system is to be subjected to system test and evaluation, an additional set may have to be produced for the end items employed in those tests. The product articles manufactured for item qualification will commonly be produced in an engineering lab by very experienced engineers and technicians, using some preliminary form of manufacturing procedures and acceptance test applied. The system test and evaluation units will commonly be produced on a low–rate production line followed by application of the planned acceptance procedure. The first production article would be subjected to acceptance, followed by a physical configuration audit (PCA), and subsequent units subjected only to acceptance. In these programs the item qualification units and those surviving DT&E should not be delivered to the customer with an expectation of full value and life; they can provide continuing program value as will be discussed under item qualification and system test and evaluation parts.

1.3.4.3.2 Low-Volume, High-Dollar Production Program

If only a very few articles are to be produced and their cost is extreme, it may not be economically feasible to produce a dedicated set of engineering items. In this case, the first article through production can be subjected to a qualification test with environmental stresses reduced somewhat from those commonly applied with dedicated engineering units. The qualification units are then refurbished to ensure they have full anticipated life and are subjected to normal acceptance test. The qualification and refurbishment steps will commonly cause the number one unit to be delivered at the end of the production run of a few units. These other units would only be subjected to a normal acceptance test. The first unit, delivered last following possible refurbishment to restore expected life, may be used as spares awaiting return of the repaired production article. In the event one of the refurbished articles must actually be used operationally due to an adverse relationship between a user operational need and the failed article repair schedule, a risk analysis would have to be accomplished to determine the degree of success that could be expected with operational use of the refurbished article. This is a common pattern for costly space instruments.

1.3.4.3.3 One-of-a-Kind Production Program

There are many kinds of programs where only one article is produced. Some examples are development and delivery of (1) a special-purpose factory, (2) a new roller coaster ride for a theme park, and (3) a new facility to dispose of or store highly dangerous nuclear waste.

Obviously, it would be uneconomical to produce a dedicated engineering system in any of these cases. Therefore, we will have to combine a system test and evaluation and acceptance process that will be applied at the system level. The system may be composed of lower tier elements that are replicated many times, leading to vendor programs that could perform qualification on dedicated units. Subsequently, produced articles would be passed through acceptance and shipped to the assembly site for installation.

Mr. Bob Grimm, a Westinghouse system engineer at the Savanna River Site, South Carolina, offered an interesting insight to the class and the author during a verification class at his site in February 2000 regarding these kinds of systems. There is actually a dual-track verification process at work in these kinds of systems. The site development agent should verify that the system (which may be a facility) satisfies the requirements for which it was designed through some form of system test and evaluation, combined with acceptance criteria. Once the site is turned over to the user, the user will have to apply an acceptance test process to the operation of the site (akin to the OT&E discussed in this book). If the site were a manufacturing site, the end product remaining within quality assurance control levels might be an adequate approach for recurring acceptance of the plant. This same approach will work for material processing plants like those used for petrochemical processing, nuclear waste processing, and food producing. In the case of a newly designed ride at an entertainment park, the recurring acceptance criteria might be focused on the safety and resulting delight of the riders. All of these cases entail an input-process-output relationship and some kind of criteria can be formulated to address output characteristics.

Another Westinghouse system engineer in the same Savanna River class, Mr. Sharad Shetty, offered that he viewed this whole verification process within a development or implementation activity in terms of input-process-output at all levels and for all applications. Given some task that is performed by machines, humans, or some combination of these, we should be interested in the quality of the results, that is, the output. Mr. Shetty expressed this in the context of his work on the job as adding adequate value to the inputs for his individual task on a program. This outlook corresponds to the U.S. Air Force interest in identifying task exit criteria for all contract tasks so that it can be logically determined when a task is complete and the quality of the value added. We can aggregate this attitude throughout all of the layers of a system development process and express the goals of verification very effectively. At the very least, a customer wants assurance that they received what they paid for in the final product. At all tiers of the work that produce that final effect, every engineer should want to know that they have contributed good value to the overall process.

1.3.4.4 Development Environment Integration

Figure 13 is an attempt to identify every conceivable development environment from which one could select the desired environment for a particular development activity. It does so through a three-dimensional Venn diagram showing combinations of different sequences (waterfall, spiral, and V), different phasing possibilities (rapid prototyping versus rigorous phasing), and one of three development attitudes (grand design, incremental, and evolutionary). The possibilities become even more numerous if we accept that three different delivery possibilities exist (high rate production, low volume/high cost, and one of a kind).

In the grand design approach, the team develops the product in a straight through process from beginning to end in accordance with a well-defined,

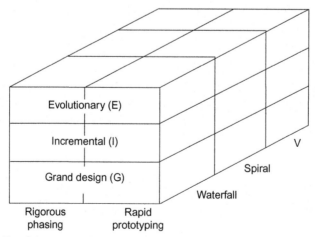

Figure 13 Development environment spaces.

predetermined plan. This attitude is normally well matched to the waterfall model.

In the incremental approach, the final requirements are defined before design work begins, but foreseeable problems, such as immature technology, prevent a straight-line approach to solution. Therefore, a series of two or more builds are planned, where the first satisfies the most pressing requirements for the product. Each incremental build permits the developer and customer to gain experience with the problem and refinement of the next design cycle, ever working toward the final design configuration. This approach is clearly coordinated with the spiral model.

In the evolutionary approach, we may be unclear about the requirements for the final configuration and conclude that we can only acquire sure knowledge of our requirements through experience with the product. This is, of course, a chicken-and-egg kind of problem, or what others might call a bootstrap problem. In order to understand the problem space, we have to experience the solution space. The overall evolutionary program is structured in a series of builds, and each build helps us to understand the final requirements more clearly as well as a more fitting design solution. Note the unique difference between the incremental and evolutionary approaches. In the incremental, we understand the final requirements and conclude that we cannot get there in one build (grand design). In the evolutionary case, we do not understand the final requirements and need experience with some design solution to help us understand them.

Earlier we said that an enterprise should have a generic process used to develop product and should repeat that process with incremental improvements as one element of their continuous improvement method. We could choose to close our minds to flexibility and interpret this to mean that we should pick one of the spaces of Figure 13 and apply only that environment in all possible situations. Alternatively, we can choose to allow our programs and development teams to apply an environment most suited to the product and development situation as a function of their unlimited choice, employing an organizational structure similar to the U.S. federal government relative to its states allowing programs some choice in selecting the precise model to be employed.

Figure 13 exposes us to a degree of complexity that we may not be comfortable with. We may be much happier with our ignorance and prefer relying on our historical work patterns. Some companies will reach this conclusion and find themselves in great trouble later. If there is anything we know about the future, it is that the pace of change will increase and the problems we must deal with will become more complex. A healthy firm interested in its future will evolve a capability that blends a generic process encouraging repetition of a standard process with a degree of flexibility with respect to allowable development environments.

These goals are not in conflict. In all of the environments discussed, we rely on requirements analysis as a prerequisite to design. In some cases, we know we may not be successful in one pass at the requirements analysis process, but the same techniques described in Chapter 2 for a functional modeling approach or for any of the other UADF will be effective in waterfall, spiral, or V development. No matter the combination of models applied, the resultant product should be verified to determine to what degree the synthesized product satisfies the requirements that drove the design process.

1.3.4.5 View from a Product Type Perspective

Systems consist of things that work together through interfaces to achieve functions. We can partition all of these things into three types for the purpose of emphasizing that we have covered all of the bases. One type of entity found in systems is hardware and the reader will find this book comprehensive in its coverage of verification as it applies to hardware. A second type of system entity is software. It is true that software development applies some unique approaches relative to hardware, because software does not really involve any physical reality. Software deals with strings of symbols forming words that are collected into programs that are run on computing machines to manipulate other strings of symbols representing information of interest. But we can identify software entities in exactly the same way we identify hardware entities in terms of our product entity structure. We prepare specifications about these entities just as we do for the hardware entities. The software entities can be verified in precisely the same pattern as described in this book. We can treat software entities at the part level purchased from suppliers in accordance with the content of the equivalent of a part specification. At whatever level of software indenture that corresponds to a software performance and detail specification, we can subject the software entity to item qualification and item acceptance. It is possible that we could deliver a software system to be operated on some particular combination of hardware computer equipment and that could be subjected to system test and evaluation against its system specification. In all of these cases we would have to partition all verification work into some number of tasks, for each of which we would have prepared a plan, procedure, and report.

The "people doing things" kind of product will be captured in technical data and as noted previously will be subjected to validation and verification in accordance with practices that are well developed and long respected.

1.4 OVERVIEW OF THE REST OF THE BOOK

1.4.1 Specifications and the Process of Creating Them

Chapter 2 describes the kinds of specifications that drive verification work, providing insight into their structure and content. The book encourages the use of a structure derived from MIL-STD-961E tailored to recognize the modeling approach selected. It is very important how the content of these specifications is obtained and Chapter 3 describes four modeling methods that can be effectively employed as the basis for every kind of specification employed on any program. This chapter applies primarily to Section 3 of specifications but also covers the whole structure. Chapter 4 provides guidance on the development of Section 4 of a specification that deals with verification. This book includes coverage of the specifications because the only hope a program has in accomplishing verification work well is driven by the quality of the specifications it creates.

1.4.2 Verification Process Design

The verification process prepared for a program should be driven by the content of the specifications prepared to cover the product entities comprising the system. An enterprise should prepare a set of documents that can be provided to each program, which can be used like templates for the development of program plans, procedures, and reports that will standardize the process employed on each program, providing support for a continuing educational experience for personnel assigned to do the verification work. Chapter 5 covers the supporting infrastructure that the enterprise should provide its programs to encourage success. Chapter 6 provides guidance for planning program verification work in general, and Chapters 7 through 10 give details for the four kinds of verification work.

1.4.3 The Four Verification Processes

In this book we have discussed four kinds of specifications and described appropriate verification practices relative to each kind. Chapter 7 deals with parts, materials, and processes. Chapter 8 covers item and interface qualification verification practices related to item and interface performance specifications. Chapter 9 covers item and interface acceptance verification practices related to the content of item and interface detail specifications. Chapter 10 deals with system test and evaluation. These four chapters form the core of the book, around which the other chapters provide supporting information.

1.4.4 Process Verification

Sadly, few enterprises apply the verification ideas applicable to determining the extent to which products they create comply with the content of program specifications to determining the extent to which their programs comply with enterprise plans, procedures, and contract provisions. Chapter 11 gives the benefits of doing so and tells how to do this important work affordably.

1.4.5 Closure

The story of verification is completed in Chapter 12. In the reader's process of arriving there, the author hopes that the reader has discovered the secret to verification success on programs. It is really very simple, but seldom realized in the author's experience.

CHAPTER 2

Specifications and Their Content

2.1 OVERVIEW OF THIS CHAPTER

Section 2.2 identifies the different kinds of specifications that have to be published on programs, and Sections 2.3, 2.4, 2.5, and 2.6 cover item and system performance specifications, item detail specifications, interface, and parts, materials, and processes (PMP) specifications, respectively, which drive the four classes of verification work (system test and evaluation and item and interface qualification, item and interface acceptance, and PMP verification). Section 2.7 discusses a series of documents that an enterprise should possess that can be provided to every new program to guide the development of specifications. Finally, the topics of paragraph numbers in specifications and requirement IDs are covered in Section 2.8. The general structure offered by MIL-STD-961E is supported with some tailoring of specification Section 3 to coordinate the structure with modeling methods selected and Section 6 to collect all traceability relationships, including verification.

2.2 KINDS OF PROGRAM SPECIFICATIONS

A specification defines all of the essential characteristics for an entity that must be designed and manufactured or purchased. These essential characteristics are referred to in this book as product requirements that the product must comply with. A specification also contains what are called verification requirements, coordinated with the product requirements, which apply to the process through which it shall be determined the extent to which the product complies with the product requirements. The product requirements, described in a companion book *System Requirements Analysis*, also published by Elsevier, identify the essential characteristics of the product design while the verification requirements identify the essential characteristics of the verification process through which it is determined the extent to which the design complies with the product requirements.

Completion of the verification requirements work falls towards the end of the requirements analysis work for every specification, occurring prior to

System Verification
http://dx.doi.org/10.1016/B978-0-12-804221-2.00002-4

the specification being reviewed, approved, placed under configuration control, and published. The verification requirements and a verification traceability matrix appearing in each specification are the principal inputs to the verification work covered in this book. Please understand that the work described in this chapter should be concluded prior to the beginning of the verification planning work relative to any product or interface entity in a system.

DoD identifies two kinds of system, item, and interface specifications: (1) performance, and (2) detail. All system specifications are of the performance kind and do not have a detail expression. Product item specifications come in both kinds with the performance kind prepared as a basis for design and item qualification planning and the detail kind prepared subsequent to design as a basis for item acceptance inspection that takes place during or subsequent to manufacturing of each product article. Interface specifications follow the pattern of item specifications. The author would argue that parts, materials, and processes specifications are only distributed to customers as detail specifications though the organization creating the PMP entity may prepare sales literature describing performance characteristics.

Earlier in DoD development programs, what are now called performance specifications were referred to as performance or Part I specifications, and detail specifications were referred to as product or Part II specifications. Figure 14 illustrates a breakdown of all the kinds of specifications we shall deal with in this book.

An applicable document shown in Figure 14 is a document referred to in a program specification as applicable to the design of the entity covered in a program specification in a particular fashion defined in a requirement. All such references are collected in a specification in Section 2 titled Applicable Documents, which is essentially a bibliography for the specification. The content of these references is every bit as much to be respected as the basic content of the specification within the boundaries in which they are referenced. Some people preparing specifications will improperly list applicable documents in Section 2 that are not called in any paragraph in the specification, as well as call an applicable document in a paragraph in Section 3 without listing it in Section 2. These are both errors to be avoided. It is also appropriate to tailor an applicable document in Section 2 so as to avoid a design being forced to completely comply unnecessarily with the full content of the document. The requirements in an applicable document commonly cover all applications, whereas a program is focused on a specific product situation.

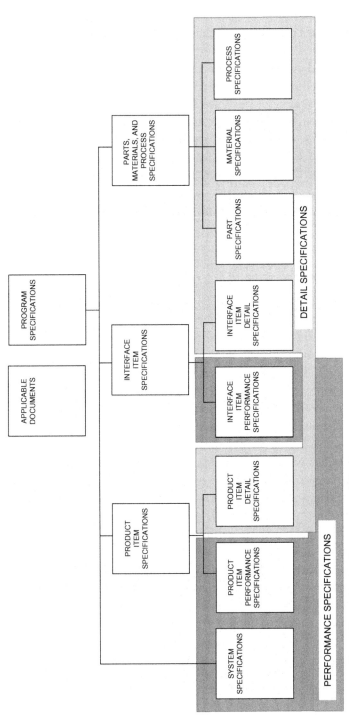

Figure 14 The different kinds of program specifications.

Product entity requirements for inclusion in the item performance specification are defined as a prerequisite to the design of the item and are written for the specific purpose of constraining the solution space so as to assure that the item designed in accordance with them will function in a synergistic way with respect to other items in a system, to achieve the system function, purpose, or need. The structure of a performance specification may follow any one of several formats, but a very well-established and effective format, recognized by the U.S. Department of Defense (DoD), entails six sections. The requirements for an item in this format are documented in Section 3, titled "Requirements." Section 4, titled "Verification," should define verification requirements for all of the product requirements defined in Section 3, give the method of verification, and system level at which verification activity shall take place in each case.

The content of the item performance specifications will drive the item qualification verification work while the system specification content will drive the system test and evaluation process. However, the content of this chapter actually applies to the verification requirements that appear in both kinds of performance specifications (item and interface) and will not be repeated in Chapter 10 of this book, dealing with system test and evaluation. There are significant differences in the development of appropriate verification requirements content for detail specifications that will be covered in Chapter 9 of the book.

Qualification verification requirements define the essential characteristics of the verification plans and procedures necessary to produce truthful evidence about whether or not the item product requirements have been fully complied with in the design. So, verification requirements do not apply to the product but rather to the process of establishing with precision and truthfulness the extent to which the design satisfies the content of Section 3 of the specification. For each requirement in Section 3 of a DoD specification, we should find one or more verification requirements in Section 4. It is good practice to also include in the specification a verification traceability matrix that establishes the relationship between Sections 3 and 4 content and states the methods that will be used to accomplish the indicated verification actions. If Section 3 includes a weight requirement, such as "3.7.2 Weight. Item shall weigh less than or equal to 134 pounds," Section 4 could contain a verification requirement such as "4.3.5 Weight. Item weight shall be determined by a scale, the calibration for which is current, that has an accuracy of plus or minus 3 ounces."

MIL-STD-490A, the military standard for program-peculiar specifications for several decades until 1995, named this Section 4 "Quality Assurance Provisions." The replacement for 490A, Appendix A of MIL-STD-961D, referred to this section as "Verification," reinforcing the author's selection of the meaning of the V words included in Chapter 1 and assumed throughout this book. The MIL-STD-961E revision continued this same structure and terminology. This book and the author accept the six-section DoD format as follows:

1. Scope
2. Applicable Documents
3. Requirements
4. Verification
5. Preparation for Delivery
6. Notes

The majority of requirements that should be verified will appear in Section 3. But, it is possible that Section 5 could include requirements that should be verified, although MIL-STD-961E encouraged that the content of Section 5 is better covered in the contract leaving Section 5 a void. The author is not supportive of this rule, especially where the developer is preparing a procurement specification, unless the developer contracts department prefers to also cover shipping requirements in the contract. A specification may include appended data that may also require verification. An example of this is where the specification contains classified material that is grouped in an appendix that can be treated as a separate document under the rules for handling classified material. The majority of the specification, being unclassified, can then be handled free of these constraints.

Section 2 of a specification includes references to other documents that are identified in specification Sections 3, 4, and 5 either requiring compliance or cited as guides for the design. This is essentially a bibliography for the specification and should not include any references that have not been called in requirements paragraphs elsewhere in the specification. Therefore, it is unnecessary to verify the content of Section 2 directly. Rather, the requirements that contain these references in specifications Sections 3 and 5 must be verified so as to embrace the content of these documents where compliance with the documents is required. The way that the document is called in the body of the specification must be considered in determining verification needs because it may limit the content to method 2 or schedule 5 whatever they happen to be, for example, or otherwise limit the applicability. Specification Section 2 may include tailoring that must be considered as well.

Generally, this tailoring will make the applicable document content less restricting. Specification Sections 1 and 6, in this structure, should contain no requirements. Section 4 can contain references to applicable documents that are therefore listed in Section 2 as well, but these calls are generally related to how certain testing and analysis activities shall be conducted.

It should be said that not all organizations capture the verification requirements in the specification covering the item they concern. The author encourages that the verification requirements be captured in the same specification as the product entity requirements, because it brings all of the item requirements into one package through which they can be managed cleanly. As we will see in this chapter, there are other places to capture them, including not at all if one wishes to encourage program failure. Failure to identify how it will be verified that the product satisfies its requirements will often lead to cost, schedule, and performance risks that we gain insight into only late in the program, when there may not be sufficient time and money to deal with them. It is a fact proven on many programs that the later in a program a fault is recognized, the more it will cost in dollars and time to correct it. Alternatively, these failures may require so much cost to fix that it is simply not feasible to correct them, forcing acceptance of lower-than-required performance and an unhappy customer. Timely application of the verification process will help to unearth program problems as early as possible and encourage their resolution, while it is still possible to do so within the context of reasonable cost constraints.

An additional reason to coordinate the preparation of specification Sections 3 and 4 is that the content of Section 3 will be better than if the two sections are crafted independently. It is very difficult to write a verification requirement if the corresponding Section 3 requirement is poorly written. This will lead to an improvement in the Section 3 requirement clarity to make it easier to write the Section 4 requirement. If one person is responsible for writing the Section 3 requirements and someone else must then craft Section 4, or the equivalent content included elsewhere at a later date, it removes a significant motivation for quality work from those responsible for Section 3.

Ideally, the same person who writes the Section 3 requirement would be required to write the corresponding verification requirements that same day. If they have trouble doing so, they will be drawn to the possibility that they have just written an unverifiable requirement, improve it, knock out the verification requirement, and move on to writing another requirement. The author has been drawn to this position through observing too frequently that separate preparation of verification requirements or the out-right rejection of the need to do so results in substantial risks that are discovered so late in the program that they are very costly to deal with at a time when there is no money left to do so. Too often programs discover the quality of their early system engineering work in item qualification, and the picture is not a happy one. In one consulting job the author was asked by a legal firm to determine the quality of the system engineering work its client had done on a program that was way behind schedule and way over cost, to protect its client from a possible government claim. The author found the verification work was a mess because the preceding specification work was badly done.

2.3 PERFORMANCE SPECIFICATION STRUCTURE

The performance specification type includes system, item, and interface performance specifications. All of the specification types shown in Figure 13 share the six-section structure listed above but let us first focus on the structure of performance specifications. All six sections have important content but three sections are especially noteworthy relative to verification. They are Sections 3, 4, and 6.

2.3.1 Performance Specification Section 3 Structure

MIL-STD-961E does offer a structure for performance specification Section 3, of course, but the author believes it to be a paragraphing structure evolved from typewriter technology days and that all of the product requirements appearing in Section 3 should be derived through modeling as explained in Chapter 3 of this book. The author would therefore organize specification Section 3 to relate to the four kinds of modeling techniques that are useful in doing so in each of the four universal architecture description frameworks (UADFs) explained briefly in Chapter 3, from which an enterprise should select their preferred modeling approach to employ on all programs. Figure 15 shows a preferred Section 3 template motivated by a clear relationship between the methods by which requirements will be derived and the location in the specification in which they will appear. This same structure will work no matter which of the four UADFs the enterprise selects to employ on every program.

In this template Paragraph 3.1 is used to explain the modeling basis for deriving requirements in each of the major portions of the section. Section 3.2 simply lists the performance requirements in Sections 3.2.1

> 3. REQUIREMENTS
> 3.1 Requirements Derivation
> 3.2 Entity Capability Requirements
> 3.3 Interface Requirements
> 3.4 Specialty Engineering Requirements
> 3.5 Environmental Requirements
> 3.6 Precedence and Criticality of Requirements

Figure 15 Performance specification Section 3 template.

through 3.2.N, each derived from the functional or behavioral modeling elements. Section 3.3 coordinates paragraph numbers with the interfaces identified on an interface-depicting figure. Section 3.4 numbers paragraphs for the specialty discipline models found to be necessary for the entity. Some of the paragraphs may require one or more layers of paragraphing. Section 3.5 includes subparagraphs for the major environmental subsets with related requirements captured in those structures derived from a trio of models. Paragraph 3.6 is a holdover from MIL-STD-961E offering an opportunity to state a program policy about precedence and criticality, including the possibility that they all share the same precedence. A customer's attitude about the content of this paragraph can be very helpful during early design work to support a trade study to determine the best combination of design features. If one configuration comes up short on meeting a requirement that is noted as less critical than others, it could still be a winner overall.

2.3.2 Performance Specification Section 4 Structure

As in the case of specification Section 3 just discussed, let us first consider an appropriate structure for Section 4 for a performance specification that includes system, item, and interface performance specifications. It is very important to recognize that the verification process appropriate to qualification driven by the content of performance specifications and acceptance driven by the content of the detail specifications has one very different characteristic. Qualification is only accomplished once on the design, so the article used in testing need only survive one inspection series and after passing that series of inspections it may be used to support further testing to discover design margins that may result in failure of the unit under test. At the very least, this testing will remove life from the unit. Each production article will have to offer the customer full life subsequent to passing acceptance testing.

So, as a minimum, even if the content of the performance and detail specification for a particular entity had a great deal in common, the verification requirements would be different to avoid overstressing production articles.

2.3.2.1 MIL-STD-961E Structure

Figure 16 gives a suggested verification section structure based on MILSTD-961E. The author hastens to mention that Figures 15 and 16 include his preferred paragraphing style, involving no title underlining, capitalization of initial letters of all words in the title, and no period at the end of the title, with the title and paragraph number on their own line followed by a blank line and the first sentence of the text beginning on the next line. The formatting encouraged by MIL-STD-961E calls for capitalization only of the first word of the title followed by a period at the title end and the text of the paragraph beginning on the same line as the title. These are small things, and you are encouraged to use whatever format your customer base prefers, unless there are conflicts across all of the members of your customer base. In this case, you would do well to find out how to serve all of your customers using a common formatting.

The term *cross-reference* is used in Paragraph 4.1.1 of Figure 16. Every specification should include what the author prefers to refer to as a *verification traceability matrix* where these cross-references can be captured in every specification. The author's preferred physical location for this matrix is in Section 6, Notes, with other traceability matrices, though some people prefer to include it in Section 4 – more on this later.

Specification Paragraph 4.1 in Figure 16 offers a place to capture information about the several classes of verification actions but one could argue that this information should be limited, as it applies to the three kinds of specifications: system specification explaining system test and evaluation; item and interface performance specifications explaining qualification, item and interface detail specifications explaining item acceptance; and PMP specifications explaining PMP verification. Paragraph 4.2 is used to define the methods of verification. It is likely that it will be necessary to partition of all of the requirements in a single specification into several separate tasks of each method and that all four methods will be required for any one specification. The special method seldom is called and the "none" method need not enter into the discussion.

Paragraph 4.3 is used to define any general verification requirements in terms of conditions in effect preceding, and/or during, and/or subsequent to the item verification series of tasks or this information has to be spread across

4. VERIFICATION

Section 4 shall include all inspections to be performed to determine that the item to be offered for acceptance conforms to the requirements in Section 3 of the specification. Verification may be accomplished by any of the methods listed in Paragraph 4.2 or any combination thereof. This section shall not include quality requirements that belong in the contract, such as responsibility for inspection, establishment of quality or inspection program requirements, warranties, instructions for nonconforming items, and contractor liability for nonconformance.

4.1 Classification of Inspections

Three classes of verification are recognized teamed with the three kinds of specifications: (1) System Test and Evaluation coordinated with the System Specification, (2) Item Qualification coordinated with the Item Performance Specifications, and Item Acceptance coordinated with item Detail Specifications.

The verification of system compliance with the content of a system specification is generally accomplished in system test and evaluation, and those inspections fall under the heading of qualifications and will be covered in Section 4 of the system specification. If Section 4 of the specification includes more than one type of inspection, a classification of inspections shall be included in this paragraph.

4.1.1 Qualification Inspection

The requirements in an item performance (or system) specification require design qualification to prove that a design complies with the content of the performance specification. Section 4 shall include a description of the inspection procedure, sequence of inspections, number of units to be inspected, and the criteria for determining conformance to the qualification requirement. It is recommended that a table be included in Section 6 that cross-references the requirements with the appropriate qualification methods and requirements. In general, a specification that has first article inspection shall not also have qualification inspection, unless it can be shown that the item is so critical that failure would likely result in death or injury.

4.1.2 Acceptance Inspection

Two forms of acceptance inspection are provided for.

4.1.2.1 First Article Inspection

When Section 4 of a detail specification specifies a first article inspection, Section 4 shall include a description of the inspection procedure, sequence of the inspections, number of units to be inspected, and the criteria for determining conformance to the requirement specified. It is recommended that a table be included in Section 6 that cross-references the requirements with the appropriate methods and first article inspection requirements.

4.1.2.2 Conformance Inspection

Conformance inspection shall ensure that production items meet detail specification requirements prior to acceptance by the Government. This process may be accomplished by a series of inspections distributed through the production process or a single post-production inspection or a sampling procedure that does not apply to every article. Conformance inspection shall include a description of the inspection procedure, sequence of inspections, number of units to be inspected, and the criteria for determining conformance to the requirement specified. Conformance examinations and tests may be the same as those specified for first article inspection, but they shall not duplicate any long-term or special tests that were used to justify inclusion of qualification in a specification. It is recommended that a table be included in Section 6 that cross-references the requirements to the appropriate methods and conformance inspection requirements.

Figure 16 Performance specification verification section structure.

4.1.3 Combined Inspections

An item specification may be published as a stand-alone performance specification in which case it will contain only qualification verification requirements in Section 4. Subsequently, a separate detail specification may be published containing only first article and/or compliance verification requirements. A third possibility is that an item specification is first published as a performance specification and later edited to include the detail specification content while continuing to contain the performance specification content. In these two cases the specification may contain both qualification and acceptance verification content.

4.2 Methods of Verification

Methods utilized to accomplish verification are defined below. The letters parenthetically included after each method appear in the verification traceability matrix in Section 6, to designate the methods required.

4.2.1 Analysis (A)

An element of verification that utilizes established technical or mathematical models or simulations, algorithms, charts, graphics, circuit diagrams, or other scientific principles and procedures in combination with human analytical thought processes to provide evidence that stated requirements were met.

4.2.2 Demonstration (D)

An element of verification that generally denotes the actual operation, adjustment, or reconfiguration of items to provide evidence that the designed functions were accomplished under specific scenarios. The items may be instrumented and qualitative limits of performance monitored.

4.2.3 Examination (E)

An element of verification and inspection consisting of investigation, without the use of special laboratory appliances or procedures, of items to determine conformance to those specified requirements which can be determined by investigations. Examination is generally nondestructive and typically includes the use of sight, hearing, smell, touch, and taste; simple physical manipulation; mechanical and electrical gauging and measurement; and other forms of investigation.

4.2.4 Test (T)

An element of verification and inspection which generally denotes the determination, by technical means, of the properties or elements of items, including functional operation, and involves the application of established scientific principles and procedures.

4.2.5 Special (S)

This method is reserved for unusual or combinations of methods including the use of modeling and simulation.

4.2.6 None (N)

Paragraphs that contain no requirements do not require verification. Paragraphs only containing a title or otherwise devoid of a characteristic that the product must possess are examples. It is generally only paragraphs in Sections 3 and 5 plus appendices that require verification.

4.3 General Inspection Requirements

4.3.1 Inspection Conditions

4.3.2 Inspection Equipment

4.4 Detailed Inspection Requirements

4.4.m Detailed Inspection Requirement m

Figure 16—cont'd

all of the item tasks in different ways. If the same equipment and conditions apply across the complete verification process, this content can be included under general inspection requirements (4.3). If the several tasks of the four methods will require different equipment, then Paragraph 4.3.2 could include a table identifying equipment versus tasks with a table partitioning requirements into tasks and another table listing and describing equipment. This same approach could be applied to conditions. Alternatively, the specification could simply list conditions and equipment required with the details of their applicability to tasks included in task plans and procedures.

Paragraph 4.4 provides the place to capture all of the verification inspection requirements numbered in any order you choose. Some engineers like to number these paragraphs by adding 4.4 to the front of the corresponding Section 3 paragraph number. This might have been a useful action back in typewriter days but today we have databases that can be used to easily capture traceability, so this kind of visual queue is not really needed. It could be helpful to capture the verification requirements under methodological subheadings, since commonly a single requirement is only going to be verified using one method and there is a chance that all of the requirements could be assigned to one task, such that the principal engineer for that task might be limited to applying a single method in his or her verification work.

2.3.2.2 An Alternative Section 4 Structure

Where the specification covers more than one class of verification, it may be useful to partition the matrix into design (or qualification), first article, and acceptance sections so that the appropriate parts can be easily referenced. Alternatively, a class column can be added to the matrix. One approach to a paragraph partitioning arrangement would be to create the paragraph structure shown in Figure 17. The author is not supportive of preparing specifications in this fashion but in some commercial situations this might be an effective and economical technique.

Section 4 can be much more simply structured so that the specification only has to refer to a single verification class, such as qualification. This would be the common case in a Part I, development, or performance specification. Similarly, a Part II, product, or detail specification can focus on the acceptance requirements. MIL-STD-961E and Figure 16 cover a structure appropriate for a specification that covers all item requirements, performance and detail. This may be common in many companies for procurement specifications but uncommon for product specifications where the item is going to be designed and manufactured by the company the program is within.

Figure 18 offers an alternate structure for a performance specification. There is only one class, qualification or design verification, so Paragraph 4.2

> 4.4 Detailed Inspection Requirements
> 4.4.1 Qualification Inspection Requirements
> 4.4.1.1 Verification Requirement 1
> 4.4.1.2 Verification Requirement 2
> 4.4.1.m Verification Requirement m
> 4.4.2 First Article Inspection Requirements
> 4.4.2.1 Verification Requirement 1
> 4.4.2.2 Verification Requirement 2
> 4.4.2.m Verification Requirement m
> 4.4.3 Acceptance Inspection Requirements
> 4.4.3.1 Verification Requirement 1
> 4.4.3.2 Verification Requirement 2
> 4.4.3.m Verification Requirement m

Figure 17 Class-differentiated Section 4 organization.

focuses only on that class. The contents of Paragraph 4.4 are the paragraph numbers that will appear in the traceability matrix, and the method entries will coordinate with the indicated paragraphing structure. The other information (test resources and equipment, for example) in the MIL-STD-961E structure could be included in Paragraph 4.3 or included in the verification planning and procedures data that we will start to address in the next chapter.

In a large specification with multiple classes applied with a lot of Section 4 content, it would be possible to apply both class and method differentiation in the paragraphing structure, combining the effects of Figures 17 and 18, but the author prefers to apply the structure in Figure 16.

2.3.3 Performance Specification Section 6 Structure

Section 6 is titled "Notes" and we may enter most anything of value to the program in this section. The example provided by Figure 19 shows only traceability and glossary data but particular specifications may stimulate those

> 4.1 Verification Methods
> 4.2 Design Verification
> 4.3 General Inspection Requirements
> 4.4 Detailed Inspection Requirements
> 4.4.1 Test Verification Requirements
> 4.4.1.1 Test Verification Requirement 1
> 4.4.1.2 Test Verification Requirement 2
> 4.4.1.m Test Verification Requirement m
> 4.4.2 Analysis Verification Requirements
> 4.4.3 Demonstration Verification Requirements
> 4.4.4 Examination Verification Requirements

Figure 18 Method-differentiated verification section structure.

6. NOTES

6.1 Requirements Traceability

6.1.1 Vertical Traceability

Vertical traceability is carried from this specification up to its parent specification. Three kinds of vertical traceability are recognized: (1) Parent-Child Traceability, (2) Source, and (3) Rationale. Upward traceability is preferred because it tends to be a one-to-one relationship whereas downward traceability tends to be one-to-many. The use of upward traceability also tends to discourage the use of flowdown as a means of determining lower tier requirements from higher tier requirements rather than depending on effective modeling that is the enterprise preferred method.

6.1.1.1 Parent-Child Traceability

Every paragraph in Section 3 shall be included in Table 6-1 linked to a paragraph in the parent specification. This table must be prepared as the requirements analysis work is in progress to prepare this child specification. It is intended that all specification content shall be derived through modeling using the enterprise preferred model set and it may develop that a particular child requirement may not possess a clear parent while possessing a clear traceability link to the modeling ID from which it was derived.

Table 6-1 Parent-Child Traceability Matrix

CHILD PARA NUMBER	CHILD PARAGRAPH TITLE	PARENT PARA NUMBER	PARENT PARAGRAPH TITLE
3.3.4.1	Mass	3.3.4.1	Mass

6.1.1.2 Source Traceability

Source traceability reveals the relationship between a requirement and the source from which it was derived. This source will most often be the supporting modeling work but may also include references to meetings, communications involving letters, phone calls, or emails, conversations, or reports. The source is captured in Table 6-2.

Table 6-2 Source and Rationale Traceability Matrix

PARAGRAPH NUMBER	SOURCE	RATIONALE
3.2.13	ERB-122, 05-12-2012, Minutes, Paragraph 3	Requirement added to clearly define the maximum permissible rate of decline in tank 2 pressure during docking maneuver.

6.1.1.3 Rationale Traceability

Rationale traceability captures the reason it was important that the requirement be identified and included in the specification. This reason is included in Table 6-2.

Figure 19 Performance specification Section 6 structure.

6.1.2 Applicable Document Internal Traceability

Table 6-3 provides the internal applicable document traceability record. All applicable documents listed in Section 2 are intended to be invoked in one or more paragraphs of this specification. If at a particular point in time in the development of this specification it is discovered that a listed document is not invoked, the comment "Not Invoked" will be carried in the INVOKING PARAGRAPH column of Table 6-3 until such time as the specification content is changed to invoke it or Section 2 is changed to delete it. If a paragraph of this specification invokes a publication not listed in Section 2 the comment "Not Listed" is included in the SECTION 2 PARAGRAPH column of Table 6-3 until such time as that document is added to Section 2 or the invoking of it has been deleted. Documents may be referred to in paragraphs of this specification for the purpose of providing information to the reader with no intention of obligating the design of the entity covered by the specification to reflect the content of those documents. In those cases the documents will not be listed in Section 2 nor will they appear in Table 6-3.

Table 6-3 Internal Applicable Document Traceability Record

SECTION 2 PARA	INVOKING PARA	DOCUMENT NUMBER	DOCUMENT TITLE
2.2.1.1	3.11.1.23.1	MIL–A–18717C	Arresting Hook Installation, Aircraft
2.2.1.2	3.6.1	MIL–S–901D (Navy)	Shock Tests, High-Impact, Shipboard Machinery, Equipment, Systems, Requirements for
2.2.2.1	Not Invoked	No Number	GSA Global Supply Catalog
2.2.3.1	3.6.17	DOD–STD–1399 Sec 070	Interface Standard for Shipboard Systems, D.C. Magnetic Field Environment

6.1.3 Longitudinal Traceability

Longitudinal traceability is maintained between the requirements in an entity specification and the related features of the design and with the related verification activity.

6.1.3.1 Design Traceability

Design traceability is maintained to encourage those responsible for the design of an entity to clearly understand the content of the specification for that entity and to consciously consider that content relative to the design decisions arrived at for the entity. It should be possible to map every requirement in Section 3 to a design feature in terms of an engineering drawing zone or a part number identified in the bill of materials. Table 6-4 provides a table containing these references.

Table 6-4 Design Traceability Matrix

PARAGRAPH NUMBER	PARAGRAPH TITLE	PARAGRAPH DESIGN REFERENCE
3.2.5.4	Altitude Hold Stability	Eng Drawing 239E2541 Sheet 5, Zone D12

Figure 19—cont'd

6.1.3.2 Verification Traceability
Verification traceability coordinates every requirement in Section 3 with a corresponding requirement in Section 4, the level at which verification shall take place (entity, parent, child), a method (test, analysis, examination, demonstration) through which the degree of compliance of the design with the requirement will be established, and an approach to be applied using that method. Subsequent action in verification planning will assign each requirement to a verification task around which all of the requirements mapped to that task will be collected. The task will be assigned to a principal engineer who will be given a list of the requirements assigned to the task. The principal will prepare a plan and procedure, accomplish the work, and prepare a report. In the report the principal must include the requirements list containing the traceability to the specific content of the report telling the extent to which the design complies.

Table 6-5 Verification Traceability Matrix

SECTION 3 PARA	PARA TITLE	SECTION 4 PARA	LEVEL	METHOD	APPROACH
3.3.5.6	Reliability	4.4.23	Entity	Analysis	Failure Rate Model

6.1.4 Modeling (Lateral) Traceability
Enterprise policy encourages that all requirements be derived from modeling. Lateral traceability maintains track between the requirements in Section 3 and the modeling artifacts from which they were derived.

Table 6-6 Lateral Traceability Matrix

PARA NUMBER	PARAGRAPH TITLE	MID	MODEL ENTITY NAME
3.2.13.2	Antenna Align	F47232	SAR, Appendix A, Sheet 32, Zone A4
3.3.5.1	Shebats Interface	I3472	SAR, Appendix C, Sheet 14, Zone D9
3.4.6.2	Reliability	H342	RAM AAA Report, Appendix F, Table 13
3.5.10.3	Rainfall	Q165	SAR, Appendix D, Sheet 213

6.1.5 System Specification Programmatic Traceability
6.1.5.1 System Specification Traceability to User Requirements
6.1.5.2 System Specification Traceability to SOW
6.2 Glossary
6.2.1 Key Words
6.2.2 Acronym List

Figure 19—cont'd

responsible for their development to include many other topics. Paragraphs 6.1.5 are only appropriate for the system specification. Provisions are included for vertical, longitudinal, and lateral Section 3 traceability. Vertical includes traceability to parent specification content as well as to the source and rationale for the requirement. Longitudinal includes design as well as verification traceability. It is seldom that engineers include design traceability in a specification and this information is only obtained by telling PDR briefers to brief what requirements design features relate to and to what extent they expect the design to be compliant. The verification table is of most importance in this book telling the linkage between Section 3 and 4 requirements, at what level it will be verified, the method to be employed, and the approach to be used with that method.

2.3.4 Performance Specification Section 2 Structure

Section 2 of a specification is essentially a bibliography containing a listing of all documents referred to in the specification, giving in each case the precise identification of these documents. There exist many broadly recognized and respected standards prepared by large customer organizations, government agencies, or industry societies. These external documents commonly take the form of standards and specifications that describe a preferred solution or constrain a solution with preferred requirements and/or values for those requirements. Often the content of these documents will coincide with the needs in a program specification. For example, the product may need a valve with characteristics that many other applications share and there has been a standard prepared for that kind of valve. Rather than preparing the detailed performance requirements for this valve, we could simply refer to the standard.

The downside is that one has to be very careful not to import unnecessary requirements through this route. If a complete document is referenced without qualification, the understanding is that the product must comply with the complete content. There are two ways to limit applicability. First, we can state the requirement such that it limits the appeal, and therefore the document applies only to the extent covered in the specification statement. The second approach is to tailor the standard using one of two techniques. The first tailoring technique is to make a legalistic list of changes to the document and include that list in the specification. The second technique is to mark up a copy of the standard and gain customer acceptance of the marked-up version. The former method is more commonly applied because it is easy to embed the results in contract language, but it can lead to a great deal of difficulty when the number of changes is large and their effect is complex.

Section 6 should include a traceability table listing the applicable documents and identifying the specification paragraph where they are called.

2.4 DETAIL SPECIFICATION STRUCTURE

The item detail specification structure is very similar to the item performance specification. The content of that structure will commonly be significantly different, however, in that the performance specification is defining requirements for the design of the product entity and the detail specification is identifying characteristics that are expected of a delivered product entity. The verification requirements will also reflect a different situation. The performance specification will call for clear proof that the design complies with requirements even to the extreme of damaging the item qualification test article, while the product completing acceptance test in response to the content of the detail specification must have a full life upon delivery.

2.5 INTERFACE SPECIFICATION STRUCTURE

The author would use the same six sections identified for a product specification as the structure for an interface specification. Section 3 of an interface performance specification can be built as a simple list of interfaces numbered 3.1, 3.2, and so forth, with each of these paragraphs defining the interface requirements in subparagraphs. Section 4 similarly can identify verification requirements for each interface. An interface detail specification may be developed as a set of engineering drawings showing the design solution to the interface defined in the performance specification. Interface requirements should not be contained in both the terminal pair of item specifications and an interface specification because it will be very difficult to maintain the data in coordination throughout the program. The best way to do this is to include the interface requirements in the interface specification and reference this document as an applicable document in the terminal pair of specifications. In the interface requirements paragraph of the item, specifications reference the interface specification listed in Section 2. If the interface in question is not covered in an interface specification, then the interface requirements must be included in the terminal pair of specifications with careful review to ensure they are compatible.

2.6 PARTS, MATERIALS, AND PROCESSES SPECIFICATIONS

The content of parts, materials, and processes (PMP) specifications depicted in Figure 13 control the characteristics of the lowest tier entities in a system. The entities in question may either be manufactured under the responsibility of the program or procured. PMP specification content should be used as the basis for verifying the entities acquired to satisfy these needs are compliant with the content of these specifications. Where PMP are procured, the verification action will often be accomplished by receiving inspection or supplier acceptance inspection. For materials, these specifications will often be single sheet spec sheets from a supplier.

2.7 SPECIFICATION GUIDANCE DOCUMENTS

The term *template* was used relative to the specification examples in Figures 15 through 18 and that term has a more precise meaning than its application to these figures. An enterprise should possess a set of templates that contain only a paragraphing structure for each of the kinds of specifications shown in Figure 14. Figures 15, 17, and 18 are actually template documents, but Figure 16 includes content in addition to paragraph structure information. Secondly, an enterprise should have a copy of a data item description (DID) for each of these documents. A DID tells how to transform a template into a specification of that kind on a program. So, a DID gives instructions on creating a program specification from an enterprise standard template. Many contractors will have customers that encourage a particular set of DIDs and they may find their customers require adherence to those DIDs, while others will find a tolerance for tailoring them. The third kinds of guidance documents that are desirable are some examples of good specifications created on past programs.

2.8 PARAGRAPH NUMBERS

We have implied that paragraph numbers should be employed to establish traceability between requirements and other things of interest. While paragraph number linkage is important for the human being, this is actually not a good practice because often paragraph numbers change over the life of a specification as material is added and removed. The engineering solution to this problem is to assign each requirement a program-wide unique requirement ID that never changes. Some requirements database tools automatically make these assignments randomly. The author prefers a base 60 ID scheme using the Arabic numerals, uppercase English alphabet letters less O, and lowercase English alphabet letters less l. Given that we take up the first place with the letter R meaning that the string is a requirement ID, and we add to that character five places, the number of unique IDs we can assign for a program would be 60^5 or 777,600,000,000. This should be quite sufficient even for a very large program. One does have to be very careful not to allow multiple cases of operator entry of paragraph numbers in the overall requirements and verification database and always to call up the unique paragraph number field, to avoid multiple paragraph numbers that can become corrupted as paragraph number changes are made over the life of the program.

CHAPTER 3

Specification Section 3 Preparation

3.1 OVERVIEW OF THIS CHAPTER

Requirements are the whole basis for verification work. They appear in program specifications. We might be inspired to inquire where these requirements come from and what their purpose might be. Requirements come in two forms: (1) requirements that define essential characteristics or properties for product or interface entities that we will refer to generically as product requirements; and (2) verification requirements that are essential characteristics of a process through which we will seek to determine the extent to which the design or manufacture of a product or interface entity complies with the product requirements contained in the specification for that entity. Requirements of both kinds appear in specifications for product and interface entities, as well as specifications for parts, materials, and processes.

Paragraph 3.2 of this book discusses the word requirement. Paragraph 3.3 provides an overview of the modeling methods that can be effectively employed to determine the content for Section 3 of a specification of any of the kinds described in Paragraph 3.2. Paragraph 3.4 covers verification requirements derived from the content of Section 3 of a specification and included in Section 4 of the specification.

Chapter 4 will form tables populated by strings consisting of paragraphs from Section 3 of every program specification paired with paragraphs in Section 4 of these specifications, the method by which verification shall be accomplished, and the level at which it shall be conducted. The important point to recognize is that the quality of the Sections 3 of these specifications predetermines in large measure the quality and affordability of the verification work that will be possible on the program.

3.2 REQUIREMENTS

This chapter summarizes the process of developing good specification content, which is covered in all of its details in a companion book by the same

System Verification
http://dx.doi.org/10.1016/B978-0-12-804221-2.00003-6
77

author and publisher titled *System Requirements Analysis.* You should try to find the edition published as an e-book in 2014. In the English-speaking world, requirements are phrased in English sentences that cannot be distinguished structurally from English sentences constructed for any other purpose. Yes, specifications are full of sentences. It is relatively easy to write sentences once you know what to write them about. A requirement is simply a statement in the chosen language that clearly defines an expectation placed on the design process prior to implementing the creative design process. Requirements are intended to constrain the solution space to solutions that will encourage small-problem solutions that synergistically work together to satisfy the large-problem (system) solution. Requirements are formed from the words and symbols of the chosen language. They include all of the standard language components arranged in the way that a good course in that language, commonly studied in the lower grades in school, specifies. Those who have difficulty writing requirements commonly experience one of four problems: (1) fundamental problem expressing themselves in the language of choice, which can be corrected by studying that language; (2) technical knowledge deficiency that makes it difficult to understand the technical aspects of the subject about which the requirements are phrased, which can be corrected through study of and gaining experience in the related technologies; (3) difficulty deciding the numerical values that should be included in requirements, which is overcome with good engineering skills and knowledge of what is possible with available technology; and (4) difficulty in deciding what to write requirements about.

The solution to the latter problem is the possession – on a personal, program, and enterprise basis – of a complete toolbox of effective requirements modeling tools and the knowledge and experience to use the tools effectively, both individually and collectively. The reader's mind probably immediately leaps to the idea of computer tools at this point, but the intent of the term *tools* deals with modeling methods, of which there are several. Computer tools can be helpful but "a fool with a tool is still a fool." If you and your organization wish to be effective in requirements work, you must master a set of modeling methods that provides you with a toolbox that can be applied, using pencil and paper or computer applications, to all programs in which the enterprise becomes involved.

Providing the toolbox should be the function of the system engineering functional department, recognizing that some departments outside of the systems engineering department will contribute people to programs, and these people will apply a department-unique method; that method should be coordinated with and be a part of the intended requirements toolbox approach. Examples of this are the reliability failure rate modeling and mass properties modeling methods applied in these two disciplines. This toolbox must encourage the identification of product characteristics that must be controlled and selection of numerical values appropriate to those characteristics. So, this toolbox must help us to understand what to write requirements about. We must accomplish requirements identification and requirements definition where we determine the values appropriate for those requirements. Requirements definition through modeling gives us a list of characteristics that we should control to encourage synergism within the evolving system definition.

If the analyst had a list of characteristics and a way to value them for specific items, language knowledge, and command of the related technologies, he or she would be in a very good position to write a specification that hits the target, which is to capture all of the essential characteristics and avoid any unnecessary content.

One approach to a solution to this problem on what to write requirements about is boilerplate, and this is essentially what specification standards such as MIL-STD-490A provided for many years – a template. Its replacement, MIL-STD-961D Appendix A and later the E Revision that integrated the content of Appendix A into the body of the document, did so for a few more years within the context of military systems. In the boilerplate approach, you have a list of paragraph numbers and titles, and you attempt to define how each title relates to the item of interest. This results in a complete requirement statement that becomes the text for the paragraph stating that requirement. The problem with this approach is that there are many kinds of requirements that cannot be specifically identified in such a generic listing. One could create a performance requirements boilerplate that covered every conceivable performance requirement category with some difficulty, only to find that it was more difficult to weed out those categories that did not apply to a specific item than it would have been to determine the appropriate categories from scratch. This is why one would find no lower level of detail in a specification standard than performance, even though 20 pages of performance requirements may evolve during the analysis for a particular specification. Interfaces are another area where boilerplate is not effective at a lower level of detail.

Many design engineers complain that their system engineers fail to flow down the customer requirements to their level of detail in a timely way. Their solution is to begin designing at their level of detail without requirements because their management has imposed a drawing release schedule they dare not violate. These engineers are often right about the failure of their system engineers, but wrong to proceed in a vacuum as if they know what they are doing. The most common failure in requirements analysis is the gap between system requirements defined in a system specification given to the contractor and the component-level requirements in the performance requirements area. In organizations in trouble, the goal in the engineering organization is often stated as getting the system performance requirements flowed down to the design level or deriving design requirements from system requirements. Unfortunately, no one in the organization seems to know quite how to do that, and avoidable errors creep into the point design solutions in that void that develops.

This and other requirements problems often observed in even good companies were the basis for the author's selection for the cover art on his *System Requirements Analysis* book published by Elsevier in early 2014. The publisher insisted on cover art and supplied several silly pictures from which to choose. The author chose a picture of three gears all enmeshed together, guaranteeing that none could turn. The author thought this expressed the problem quite well. As the author tries to show in that book, an enterprise need not fall victim to that problem.

Flowdown is a valid requirements analysis strategy, as is boilerplate (which the author refers to as one form of cloning); they are just not completely effective tools across all of the kinds of requirements that must be defined, and particularly not effective for performance requirements. Flowdown is composed of two subordinate techniques: allocation (or partitioning) and derivation. Allocation is applied where the child requirement has the same units as the parent requirement and we partition the parent value into child values in accordance with some mathematical rule. Where the child requirement has different units than the parent requirement, the value may have to be derived from one or more parent requirements through parametric analysis, use of mathematical models, or use of simulation techniques.

The problem in the flowdown strategy is that you must have a set of parent requirements in order to perform requirements analysis. One may ask, where did the parent requirements come from? This difficulty is solved by applying some form of modeling leading to the identification of characteristics that the item must have. The model should be extensible with respect to the problem represented by the system being developed, such that it can be used to identify needed functionality and be iteratively expanded for lower tiers of the problem. Functional analysis (described later in this chapter) is an example of a means to identify performance requirements. Once the physical hierarchy of the product structure is established through functional analysis and allocation, other models can be applied to identify design constraints and appropriate values in interface, environmental, and specialty engineering areas. In the latter two cases, requirements allocation can be very effective in a requirements hierarchy sense.

Our toolbox should help us understand what to write requirements about and what characteristics we should seek to control. Once we have this list of characteristics, we must pair each with an appropriate numerical value and weave words around them to make the meaning very clear. We need to quantify our requirements because we will want to be able to determine whether or not a particular design solution satisfies the requirement. This is very hard to do when the requirement is not quantified. As we shall see, this is the reason why the author encourages simultaneous definition of requirements and the companion verification requirements for determining the degree to which the solution satisfies them.

Our system requirements analysis process should clearly tell us what kinds of characteristics we should seek to write requirements about and provide a set of tools to help us add characteristics to our list. We also need to have the means to quantify them. We seek a list of characteristics that is complete, in that it does not omit any necessary characteristics, and minimized, since we wish to provide the maximum possible solution space for the designer. We would like to have a more specific understanding of how this requirements analysis solution is characterized by completeness and minimization, but there is no easy answer. The best prescription the author can offer is to apply an organized approach that connects specific models up with specific requirements categories and apply the models with skill based on knowledge derived in practice or through training.

Figure 20 offers an overall taxonomy of every kind of requirement that one would ever have to write in the form of a three-dimensional Venn diagram. The top layer, Part I in MIL-STD-490A or the performance specifications in the context of MIL-STD-961E, corresponds to development requirements, often called design-to requirements, which must be clearly understood before design work is begun. The lower layer, Part II in 490A context or detailed specifications content under 961E, corresponds to product requirements, commonly called build-to requirements. The portion of the construct below the heavy middle line is for product specifications, in which we are primarily interested in this book. The requirements

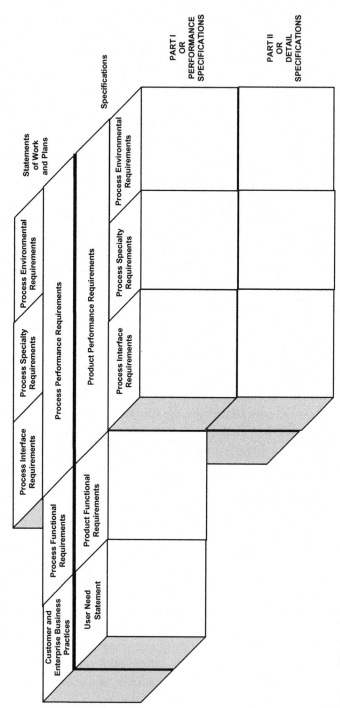

Figure 20 Requirements taxonomy.

above this line correspond to process requirements captured in statements of work and plans. Many of the same techniques discussed in this book for specifications apply to process requirements, but most of the book specifically concerns product specifications.

Development requirements can be further categorized, as shown in Figure 20, as performance requirements and design constraints of three different kinds. Performance requirements tell what an item must do and how well it must do it. Constraints form boundary conditions within which the designer must remain while satisfying the performance requirements. The toolbox encouraged in this book evolves all performance requirements from the customer need statement, the ultimate system function, through a modeling process that identifies lower tier functions from which performance requirements are derived.

These performance requirements are allocated to things in the evolving system physical model, commonly a hierarchical structure. Functional analysis provides one set of models that also includes three other sets for the three kinds of models useful in deriving three kinds of constraints.

Verification requirements are not illustrated in Figure 20 because they are paired with the item requirements in our imagination. One could construct another dimension for design and verification requirements, but it would be very hard to envision and draw in two dimensions. This solution corresponds to the fundamental notion that flows through this whole book that a design requirement should always be paired with a verification requirement at the time the design requirement is defined. The reason for this is that it results in the definition of much better design requirements when you are forced to tell how you will determine whether or not the resultant design has satisfied that design requirement. If nothing else, it will encourage quantifying requirements because you will find it very difficult to tell how they will be verified without a numerical value against which you can measure product characteristics.

Before moving on to the toolbox for the other categories illustrated in Figure 20, we must agree on the right timing for requirements analysis. Some designers reacted to the rebirth of the concurrent design approach with acceptance, because they believed it meant that it was finally okay again to simultaneously develop requirements and designs. That is not the meaning at all, of course. The systems approach, and the concurrent development approach which has added some new vocabulary to an old idea, seek to develop all of the appropriate requirements for an item prior to the commitment to design work. We team the many specialists together to first understand the problem phrased in a list of requirements. We then team together to create a solution to that problem in the form of a product design and coordinated process (material, manufacturing, quality, test, and logistics) design. The concurrency relates to the activity within each fundamental step in the systematic development of product accomplished in three serial steps (requirements, synthesis, and verification), not to all of these steps simultaneously. Even in the spiral sequence, we rotate between these steps serially repeatedly.

It is true that we may be forced by circumstances to pursue design solutions prior to fully developing the requirements (identifying the problem), as discussed earlier under development environments, but requirements before design is the intent. A case where this is appropriate is when we wish to solve a problem before we have sufficient knowledge about the problem and it is too hard to define the requirements in our ignorance. A physical model created in a rapid prototyping activity can provide great insights into appropriate requirements that would be very difficult to recognize through a pure structured approach. This is what we do in a spiral model, of course, but it really is a case of requirements before design repeated, as we gain a better understanding of the requirements in the application of each spiral.

3.3 FOUR COMPREHENSIVE MODELS

We can separate the models of value in supporting architecture and requirements engineering work into three sets. The first one will be referred to as problem space models, which are effective in analyzing the system problem space that can be described in terms of the functionality or behavior needed to resolve the problem represented by the customer's need statement. The performance requirements flow out of this model. The second set is useful in capturing the product and interface relationship structure of the system to be developed. Both the product entities and the relationships between them are exposed in the problem space model driven by the way functionality and performance requirements derived from them are coordinated with product entities. The third set will expose us to appropriate requirements in system constraint areas in terms of interfaces, specialty engineering, and environmental stresses.

Requirements are most effectively derived using models of the problem space a program must deal with. Over the past 50 or more years, enterprises developing systems have for the most part failed to focus on a single comprehensive model and insist that their employees become proficient in its use. One of the reasons for this failure is that no one has come forward with a clear identification of one of these models. There have been devotees of models that are effective for systems and hardware, and those effective for software. The author has identified four of these comprehensive universal architecture description frameworks (UADFs) in a paper titled *Universal Architecture Description Framework* appearing in Volume 12 Number 2 Summer 2009 edition of *Systems Engineering: The Journal of The International Council on Systems Engineering*, any one of which can be effectively used as the basis for deriving all requirements for a system and its component entities, no matter how it is determined how the design will be implemented in terms of hardware, software, or people doing things.

In the companion book, *System Requirements Analysis*, these four universal architecture description frameworks are also covered in some detail: (1) functional, (2) modern structured analysis combined with process for system architecture and requirements engineering (MSA-PSARE), (3) unified modeling language combined with system modeling language (UML-SysML), and (4) a fourth is added in the form of the unified process for DoDAF-MoDAF (UPDM). Other than the functional model, which was developed before any software or digital machines on which it might run were developed, all of these models were developed for application to software development, but can also be effective for development of hardware and systems as well. This section briefly offers an overview of these four modeling approaches, plus the additional models required to complete the picture for product entity, interface, specialty engineering, and environmental requirements.

Any of these four UADFs combined with all or parts of the additional models can be applied comprehensively to identify the physical artifacts a system shall be composed of, how they shall have to be interfaced, and what the product requirements for these entities should be, no matter how you later decide to implement them. An enterprise should adopt one of the four UADFs introduced in this chapter and covered in detail in the companion *System Requirements Analysis* book, also written by the author and published by Elsevier, and apply it on all programs. The result will be the derivation of all requirements from and traceability of those requirements to models and progressive improvement of the process for doing so on programs through repetition.

3.3.1 The Functional Model

The first model was developed many years ago, was adapted to software in the form of flow charts to cover software, and has been expressed in many different ways. Its application to software development automatically pushes it into the UADF set but there are few software engineers who would volunteer to use it for that purpose today.

3.3.1.1 The Need and Its Initial Expansion Using Functional Analysis

The very first requirement for every system is the customer's need: that is, a simple statement describing the customer's expectations for the new, modified, or reengineered system. Unfortunately, the need is seldom preserved once a system specification has been prepared, and thus system engineers fail to appreciate a great truth about their profession. That truth is that the development process should be characterized by logical continuity from the need throughout all of the details of the system.

The models discussed in this section are designed to expand upon the need statement to fully characterize the problem space. The exposed functionality is allocated to things and translated into performance requirements for those things identified in the physical plane. We then define design constraints appropriate to those things. All of our verification actions will map back to these requirements. We will first discuss entering the problem space briefly defined by the need statement, using traditional structured analysis employing some form of functional flow diagram, a requirements analysis sheet in which to capture the requirements and allocate them to product entities piled up in a hierarchical product entity structure to identify the things the system is composed of, and then determine, based on the way functionality is to be implemented in the product structure, what the interfaces must be between those entities. We complete this modeling process and the requirements to be captured in specifications through the application of a set of design constraint models for specialty engineering, environmental, and interface requirements. Subsequently, we will sketch in the alternative use of the other three UADFs.

The need statement seldom provides sufficient information by itself to ignite this analytical process, so it may be followed by contractor efforts to understand the customer's need and the related mission or desired action or condition. In military systems this may take the form of a mission analysis. In commercial product developments we may seek to understand customer needs through surveys, focus groups, and conversations with selected customer representatives.

This early work is focused on two goals. First, we seek to gain expanded knowledge of the need. In the process we seek to ignite the functional decomposition process targeted on understanding the problem represented by the need and making decisions about the top-level elements of the solution by allocating performance requirements derived from top-level functionality to major items in the product entity structure. Some system engineers refer to this process as requirements analysis, but it is more properly thought of as the beginning of the system development process involving some functional analysis, some requirements analysis, and some synthesis and integration. Second, we seek to validate that it is possible to solve the problem stated in the need. We are capable of thinking of needs that are not possible to satisfy. The problem we conceive may be completely impossible to solve, or it may only be impossible with the current state of the art. In the latter case, our study may be accomplished in parallel with one or more technology demonstrations that will remove technological boundaries, permitting development to proceed beyond the study. Lockheed found that the SR-71 would fly so fast that the aluminum structure would not survive the heat generated in passage through even the thin air at high altitude, and they had to invent the process of manufacturing with titanium, enabling the program to proceed.

This early activity should include some form of functional analysis as a precursor to the identification of the principal system entities. These top-level allocations may require one or more trade studies to make a knowledge-driven selection of the preferred description of the solution. This process begins with the recognition that the ultimate function is the need statement and that we can create an ultimate functional flow diagram (one of several decomposition techniques) by simply drawing one block for this function F that circumscribes a simplified version of the need statement. The next step is to craft a life-cycle model such as Figure 21 showing a block called Use System, identified in the diagram as function F47. At the time we may not know much about what it means for the customer to use the system, but we know this function will have to exist.

It should be noted that the user community of the customer will most often state its requirements in terms of the missions they intend to accomplish. The contractor interested in becoming the selected contractor to develop the system is interested in the engineering problem they will have to solve. This same difference comes into play at the end of the development work as well, with the contractor required to prove that the system complies with the content of the system specification containing engineering requirements through development test and evaluation and the user intended to determine the extent to which the delivered system complies with the original mission-oriented user requirements.

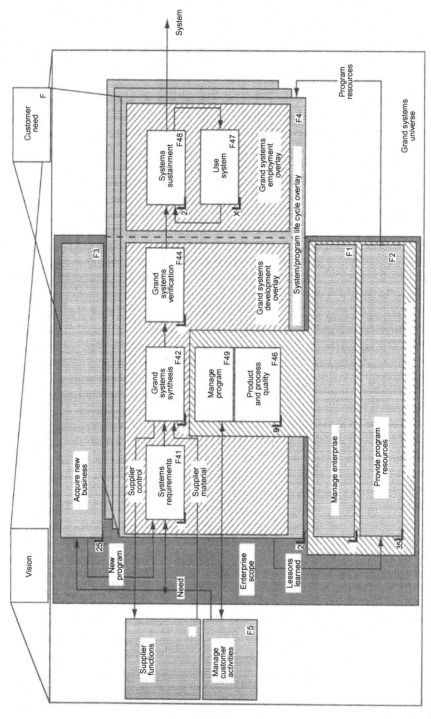

Figure 21 Functional system life-cycle model.

Figure 21 actually represents a life-cycle model for a system with main emphasis on the development of the system that is expected to accomplish the functionality contained within the function block F47 titled Use System. During function block F41 the program must come to clearly understand the customer's view of the Use System function, accomplishing a functional analysis of that function.

The reader will note that all of the function blocks in Figure 21 are identified by a modeling ID beginning with the letter F for function, followed by a series of characters. The author uses a base-60 identification system where each character may be from the set composed of upper-case English alphabet letters less the letter "O," the lower-case English alphabet letters less the letter "l," and the Arabic numerals 0 through 9. The author insists on uniquely identifying every modeling artifact such that traceability can be defined between requirements derived from the model and the specific artifacts they were derived from.

This life-cycle model will be the same for every system the enterprise ever develops. Given this expansion of the ultimate function, the system need, we seek to identify a list of functions subordinate to F47 such that the accomplishment of the related activities in some particular sequence assures that the need is satisfied. We seek to decompose the grand problem represented by the need into a series of smaller, related problems and determine what kinds of resources could be applied to solving these smaller problems in a way that their synergism solves the greater problem.

If our need is to transport assigned payloads to low Earth orbit, we may identify lower-order problems, or functions, expanding the function F47, Use System, as illustrated in Figure 22. The particular sequence of lower-tier functions illustrated is keyed to the use of some kind of chemical rocket launched from the Earth's surface. If we chose to think in terms of shooting payloads into orbit from a huge cannon, our choice of lower-tier functionality would be a little different, as it would be if we elected the Orbital Science Corporation's Pegasus solution of airborne launch.

If our need was the movement of 500,000 cubic yards of naturally deployed soil per day in 1935, we might have thought in terms of past point design solutions involving a steam shovel and dump trucks. At some point in our history, someone considered other alternatives, thus inventing the earthmover. If the material to be moved is naturally deployed rock or coal, we may need to use an explosive as a preparatory step.

So, the major functions subordinate to the need are interrelated with the mission scenario, and we cannot think of the two independently. The

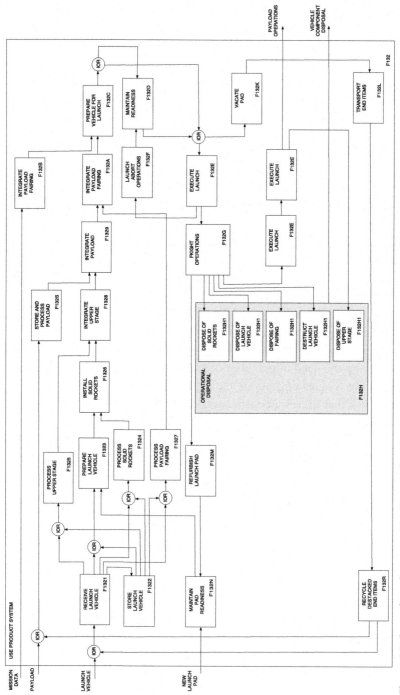

Figure 22 Space launch system Use System function.

thought process that is effective here is to brainstorm several mission scenarios and develop corresponding top-level functional flow diagrams at the highest level for these scenarios. We may then determine ways of selecting the preferred method of stating the problem by identifying key parameters that can be used to examine each scenario. This is essentially a requirements analysis activity to define quantified figures of merit useful in making selections between alternative scenarios.

As noted previously, we may have to run trade studies to select the most effective problem statement as a precursor to allocation of top-level functionality to things. We may also have to accomplish trade studies as a means to derive a solution based on facts for the most appropriate allocation of particular functions. Finally, we may have to trade the design concept for particular things conceived through the allocation process. Trade studies are but organized ways to reach a conclusion on a very difficult question that will not yield to a simple decision based on the known facts. All too often in engineering, as in life in general, we are forced to make decisions based on incomplete knowledge. When forced to do so, we should seek out a framework that encourages a thorough – no, systematic – evaluation of the possibilities, and the trade study process offers that.

The customer need statement need not be focused on a new unprecedented problem, as suggested in the preceding comments. The customer's need may express an alteration of a problem previously solved through earlier delivery of a system. The early study of this kind of problem may involve determining to what extent the elements of existing systems can be applied and to what extent they must be modified or elements replaced or upgraded. This is not a radically different situation. It only truncates the development responsibilities by including within the system things that do not require new development. Those items that do require new development will respond to the structured approach expressed in this chapter at their own level of difficulty.

This early analytical process may start with no more than an expression of the customer need in a paragraph of text. The products with which we should conclude our early mission analysis work include the following:

1. A master functional flow diagram (or equivalent diagrammatic treatment) coordinated with a planned mission scenario briefly described in text and simple cartoon-like graphics.
2. A product entity block diagram defining the physical model at the top level.
3. A top-level schematic block diagram or n-square diagram that defines what interfaces must exist between the items illustrated on the product entity block diagram.

4. Design concept sketches for the major items in the system depicted on the product entity block diagram.

5. A record of the decision-making process that led to the selection of the final system design concept with references to approved reports and other documents.

6. A list of quantified system requirement statements. These may be simply listed on a single sheet of paper or captured in a more formal system or operational requirements document, or preliminary system specification.

So, this mission analysis activity is but the front end of the system development process using all of the tools used throughout the development downstroke. It simply starts the ball rolling from a complete stop and develops the beginning of the system documentation. It also serves as the first step in the requirements validation activity. Through the accomplishment of this work, we either gain confidence that it is possible to satisfy this need or conclude that the need cannot be satisfied with the available resources at our disposal. We may conclude, based on our experience, that we should proceed with development, await successful efforts to acquire the necessary resources, focus on enabling technology development for a period of months or years, or move on to other pursuits that show more promise. In the case of commercial products, this decision process may focus on marketing possibilities based on estimates of consumer demand and cost of production and distribution.

3.3.1.2 Structured Decomposition Using Functional Analysis

Structured decomposition is a technique for decomposing large complex problems into a series of smaller related problems. We seek to do this for the reasons discussed earlier. We are interested in an organized or systematic approach for doing this because we wish to make sure we solve the right problem and solve it completely but within the constraints of available time and money. We wish to avoid, late in the development effort, finding that we failed to account for some part of the problem that forces us to spend additional time and money to correct and brings into question the validity of our current solution. We wish to avoid avoidable errors because they cost so much in time, money, and credibility. This cost rises sharply the further into the development process we move before the problem is discovered.

The understanding is that the problems we seek to solve are very difficult and that their solution will require many people, each specialized in a particular technical discipline. Further, we understand that we must encourage these specialists to work together to attain a condition of synergism of their knowledge and skill and apply that to the solution of the complex problem. This is not a field of play for rugged individuals except in the leadership of these bands of specialists. They need skilled and forceful leadership by a person possessed of great knowledge applicable to the development work and able to make sound decisions when offered the best evidence available.

During the development of several intercontinental ballistic missile (ICBM) systems by the U.S. Air Force, a very organized process called *functional analysis* came to be used as a means to thoroughly understand the problem, reach a solution that posed the maximum threat to our enemies consistent with the maximum safety for the United States, and make the best possible choices in the use of available resources. We could argue whether this process was optimum and successful in terms of the money spent and public safety, but we would have difficulty arguing with the results following the demise of the Soviet Union as a threat to the future of the United States for a considerable period of time.

At the time these systems were in development, computer and software technology were also in a state of development. The government evolved a very organized method for accumulating the information upon which development decisions were made, involving computer capture and reporting of this information. Specifications were prepared on the organized systems that contractors were required to respect for preparing this information. Generally, these systems were conceived by people with a paper mindset within an environment of immature computer capabilities. Paper forms were used based on the Hollerith 80-column card input-output and were intended as data entry forms. These completed forms went to a keypunch operator. The computer generated poorly crafted report forms. But, this was a beginning and very successful in its final results.

This process included at its heart a modeling technique called *functional flow diagramming*, discussed briefly previously. The technique uses a simple graphical image created from blocks, lines, and combinatorial flow symbols to model or illustrate needed functionality. It was no chance choice of a graphical approach to do this. It has long been well known that we humans can gain a lot more understanding from pictures than we can from text. It is true that a picture is worth 10^3 words. Imagine for a moment the amount of information we take in from a road map glanced at while driving in traffic and the data rate involved. All of the structured decomposition techniques employ graphical methods that encourage analytical completeness as well as minimizing the time required to achieve the end result.

While functional flow diagramming was an early technique useful in association with the most general kind of problem, computer software analysis has contributed many more recent variations better suited to the narrower characteristics of software. U.S. Air Force ICBM programs required adherence to a system requirements analysis standard and delivery of data prepared in accordance with a data item description for functional flow diagramming. At the time, functional flow diagramming in the form of flow charts was used as a model for software development, resulting in the use of one model for all development work.

While functional flow diagramming is still a very effective technique for grand systems and hardware, many people would argue it is not as effective for computer software development as other techniques since developed specifically for it. So, our toolbox of analytical techniques should include a modeling approach that will work well no matter how we decide to implement the solution in terms of hardware, software, and people doing things. The functional approach will work, but few software engineers would choose it today.

3.3.1.3 Functional Analysis Continuation

Functional flow diagramming has been the author's preferred approach for grand systems in the past because of its simplicity and generality and his background as a hardware-dominated system engineer. This process starts with the need as function F, which is expanded into a set of next-tier functions, which are all things that have to happen in a prescribed sequence (serial, parallel, or some combination) to result in function F being accomplished. One draws a block for each lower-tier activity and links them together in a sequence using directed line segments to show sequence. Logical OR and AND symbols are used on the connecting lines to indicate combinatorial possibilities that must be respected. This process continues to expand each function, represented by a block, into lower-tier functions. Figure 23 sketches this overall process for discussion.

A function statement begins with an action verb that acts on a noun term. The functions exposed in this process are expanded into performance requirements statements that numerically define how well the function must be performed. This step can be accomplished before allocation of the performance requirement or function statement to a thing in the product entity structure or after. But, in the preferred case, the identification of the function obligates the analyst to write one or more performance requirements derived from the function and allocate that performance requirement to an entity

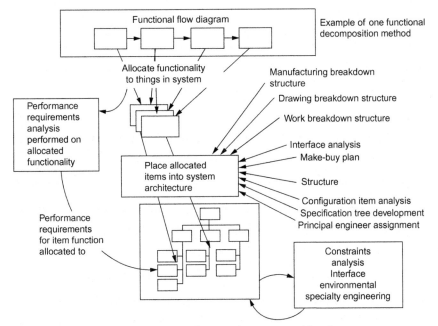

Figure 23 Structured decomposition for grand systems and hardware.

that must be designed to comply. This is the reason for the power of all modeling techniques. They are exhaustively complete when done well by experienced practitioners. It is less likely that we will have missed anything, compared to an ad hoc approach.

This process begins with the need and ends when the lowest tier of all items in the physical product entity structure satisfies one of these criteria: (1) the item will be purchased from another company at that level, or (2) the developing organization has confidence that it will surrender to detailed design by a small team within the company and that the corresponding problem is sufficiently understood in either case, so that an adequate specification can be prepared.

There are two extreme theories on the pacing of the allocation process relative to the functional decomposition work. Some system engineers prefer to remain on the problem plane as long as possible to ensure a complete understanding of the problem. This may result in a seven-tier functional flow diagram before anything is allocated. In the other extreme, the analyst expands a higher-tier function into a lower-tier diagram and immediately allocates the exposed functionality. This selection is more a matter of art and experience than science, but the author believes a happy medium between the extremes noted is optimum.

If we accumulate too much functional information before allocation, we run the risk of a disconnect between lower-tier functions and the design concepts associated with the higher-order product entities that result. If, for example, we allocate a master function for "Transport dirt" to an earth mover, we may have difficulty allocating lower-tier functionality related to moving the digging and loading device (which, in our high-level design concept, is integrated with the moving device). Allocation accomplished too rapidly can lead to instability in the design concept development process because of continuing changes in the higher-tier functional definition.

The ideal pacing involves progressive allocation. We accumulate exposed functionality to a depth that permits us to thoroughly analyze system performance at a high level, possibly even run simulations or models of system behavior under different situations with different functionality combinations and performance figures of merit values in search of the optimum configuration. Allocation of high-order functionality prior to completing these studies is premature and will generally result in a less-than-optimum system and many changes that ripple from the analysis process to the architecture synthesis and concept development work. Throughout this period we have to deal with functional requirements rather than raw function statements, so, when we do allocate this higher-order functionality, it will be as functional requirements rather than raw function names. Before allocating lower-tier functionality, we should allocate this higher-order functionality and validate it with preliminary design concepts. These design concepts should then be fed back into the lower-tier functional analysis to tune it to the current reality. Subsequent allocations can often be made using the raw functions, followed by expansion of them into performance requirements after allocation.

Some purists would claim that this is a prescription for point designs in the lower tiers. There is some danger from that, and the team must be encouraged to think through its lower-tier design concepts for innovative alternatives to the status quo. The big advantage, however, to progressive tuning of the functional flow diagram through concept feedback is that at the lower tiers the functional flow diagram takes on the characteristics of a process diagram, where the blocks map very neatly to the physical situation that the logistics support people must continue to analyze. This prevents the development of a separate logistics process diagram with possible undesirable differences from the functional flow diagram. Once again, we are maintaining process continuity.

The author believes the difference between functional flow and process diagrams is that the former is a sequence of things that must happen, whereas the latter is a model of physical reality. When we are applying the functional flow diagram to problem solving, we do not necessarily know what the physical situation is nor of what items the system will consist. The blocks of the process diagram represent actual physical situations and recognition of specific resources.

The U.S. Air Force developed a variation of functional flow or process diagramming called the IDEF diagram. IDEF is a compound acronym originally meaning "ICAM Definition," where ICAM = Integrated Computer Aided Manufacturing Analysis. Today, IDEF is more often read simply as integrated definition. In addition to the horizontal inputs and outputs that reflect sequence, these diagrams also have inputs at the top and bottom edges that reflect controlling influences and resources required for the steps, respectively. This diagrammatic technique was developed from an earlier structured analysis and design technique (SADT), a diagramming technique developed for software analysis and applied to the development of contractor manufacturing process analysis. It does permit analysis of a more complex situation than simple process diagramming, but the diagram developed runs the risk of totally confusing the user with the profusion of lines. Many of the advantages claimed for IDEF can be satisfied through the use of a simpler functional or process flow diagram teamed with some tabular data. These diagrams present a simple view that the eye and mind can use to acquire understanding of complex relationships, and a dictionary listing of the block titles presents details related to the blocks that would confuse the eye if included in the diagram. The IDEF technique has evolved into an IDEF-O for function analysis, IDEF-1X for relational data analysis, IDEF-2 for dynamic analysis, and IDEF-3 for process description.

Some system engineers, particularly in the avionics field, have found it useful to apply what can be called hierarchical functional analysis. In this technique, the analyst makes a list of the needed lower-tier functionality in support of a parent function. These functions are thought of as subordinate functions in a hierarchy rather than a sequence of functions as in flow diagramming. They are allocated to things in the evolving architecture generally in a simple one-to-one relationship. The concern with this approach is that it tends to support a leap to point design solutions familiar to the analyst. It can offer a very quick approach in a fairly standardized product line involving modular electronics equipment as a way to encourage completeness of the analysis. This technique also does not support time line analysis as does functional flow diagramming, since there is no sequence notion in the functional hierarchy.

Ascent Logic popularized another technique, called *behavioral diagramming*, that combines the functional flow diagram arranged in a vertical orientation on paper with a data or material flow arranged in the horizontal orientation. The strength of this technique is that we are forced to evaluate needed functionality and data or material needs simultaneously rather than as two separate, and possibly disconnected, analyses. The tool RDD-100, offered by Ascent Logic until its bankruptcy, used this analysis model, leading to the capability to simulate system operation and output the system functionality in several different views including functional flow, IDEF-O, or n-square diagrams. Behavioral diagramming was actually a rebirth of IPO developed for mainframe computer software development. It included a vertical flow chart and a lateral commodity flow that could be used for data (as in the original IPO) or material acted upon by the functional blocks.

Another system engineering tool company, Vitech, developed a tool called CORE that uses a similar diagramming treatment called enhanced functional flow block diagramming (EFFBD), with the functional flow done in the horizontal and the data or material flow in the vertical axis. These two techniques and IDEF-O are two-axis models that permit a much richer examination of the problem space than possible with simple functional flow diagramming, but they all suffer from diagram visual complexity that makes it more difficult for the human to move the image and its meaning into his or her mind through vision.

Whatever techniques we use to expose the needed functionality, we have to collect the allocations of the performance requirements derived from that functionality into a hierarchical product entity block diagram reflecting the progressive decomposition of the problem into a synthesis of the pre-ferred solution. The peak of this hierarchy is the block titled system, which is the solution for the problem (function) identified as the need. Subordinate elements, identified through allocation of lower-tier functionality, form branches and tiers beneath the system block. The product entity structure should be assembled recognizing several overlays to ensure that everyone on the program is recognizing the same structure, viewed as a work break-down structure (finance), manufacturing breakdown structure (manufacturing assembly sequence), engineering drawing structure, specification tree, configuration or end item identification, make-buy map, and development responsibility matrix.

As the product entity structure is assembled, the needed interfaces between these items must be examined and defined as a prerequisite to defining their requirements. These interfaces will have been predetermined by the way that we have allocated functionality to things and modified as a function of how we have organized the things in the architecture and the design concepts for those things. During the product entity synthesis and initial concept development work, the interfaces must be defined for the physical model using schematic block or n-square diagrams.

3.3.1.4 Performance Requirements Analysis

Performance requirements define what the system or item must do and how well it must do those things. Precursors of performance requirements take the form of function statements or functional requirements (quantified function statements). These should be determined as a result of a functional analysis process that decomposes the customer need, as noted previously, using an appropriate flow diagramming technique.

Many organizations find that they fail to develop the requirements needed by the design community in a timely way. They keep repeating the same cycle on each program and fail to understand their problem. This cycle consists of receipt of the customer's requirements or approval of their requirements in a specification created by the contractor, followed by a phony war on requirements, where the systems people revert to documentation specialists and the design community creates a drawing release schedule in response to management demand for progress. As the design becomes firm, the design people prepare an in-house requirements document that essentially characterizes the preexisting design. Commonly, the managers in these organizations express this problem as, "We have difficulty flowing down system requirements to the designers."

The problem is that the flowdown strategy is only effective for some specialty engineering and environmental design constraints. It is not a good strategy for interface and performance requirements. It is no wonder that these companies have difficulty. There is no one magic bullet for requirements analysis. One needs the whole toolbox described in this chapter. Performance requirements are best exposed and defined through the application of a structured process for exposing needed functionality and allocation of the exposed functionality to things in the product entity structure. You need not stop at the system level in applying this technique. It is useful throughout the hierarchy of the system.

Performance requirements are traceable to (and thus flow from) the process from which they are exposed much more effectively than in a vertical sense through the product entity structure. In the context of Figure 22, they trace to the problem or functional plane. Constraints are generally traceable within the physical or solution plane. This point is lost on many engineers and managers, and thus they find themselves repeating failed practices indefinitely.

Given that we have an effective method for identifying valid performance requirements as described previously under functional analysis, we must have a way to associate them with quantitative values. In cases where flowdown is effective, within a single requirement category, such as weight, reliability, or cost, a lower-tier requirement value can be determined by allocating the parent item value to all its child items in the product entity structure. This process can be followed in each discipline, creating a value model for the discipline. Mass properties and reliability math models are examples. In the case of performance requirements, we commonly do not have this clean relationship, so allocation is not always effective in the same way.

Often the values for several requirements are linked into a best compromise, and to understand a good combination we must evaluate several combinations and observe the effect on selected system figures of merit such as cost, maximum flight speed of an aircraft, automobile operating economy, or land fighting vehicle survivability. This process can best and most quickly be accomplished through a simulation of system performance where we are allowed to control certain independent parameters and observe the effects on dependent variables used to base a selection upon. We select the combination of values of the independent variables that produces the best combination of effects in the dependent variables.

Budget accounts can also be used effectively to help establish sound values for performance requirements. For example, given a need to communicate across 150 miles between the Earth's surface and a satellite in low Earth orbit, we may allocate gain (and loss) across this distance, the antenna systems appearing at both ends, connecting cables, receiver, and transmitter. Thus the transmitter power output requirement and the receiver sensitivity requirement are determined through a system-level study of the complete communications link.

3.3.1.5 Process Requirements Analysis

The techniques appropriate to product requirements analysis may also be turned inwardly toward our development, production, quality, test, and logistics support processes. Ideally, we should be performing true cross-functional requirements analysis during the time product requirements are being developed. We should be optimizing at the true system level, involving not only all product functions but process functions as well. We should terminate this development step with a clear understanding of the product design requirements and the process design requirements. At this point we will be clear about what the system shall consist of and could switch from functional analysis to process analysis, the difference being that the former deals with functions that must be accomplished while the latter deals with physical realities.

We often make the mistake of drawing a product functional flow diagram only focused on the operational mission of the product. Our top-level diagram should recognize product development and testing, product manufacture and logistic support, and product disposition at the end of its life. This should truly be a life-cycle diagram. Figure 24 is an example of such a total process flow diagram translated from an earlier life-cycle functional flow diagram.

System development (Pl), material acquisition and disposition (P2), and integrated manufacturing and quality assurance (P3) processes can be represented by program evaluation review technique (PERT) or critical path method (CPM) diagrams using a networking and scheduling tool. The deployment process (P5) may entail a series of very difficult questions involving gaining stakeholder buy-in as well as identification of technical, product-peculiar problems reflecting back on the design of the product. At least one intercontinental ballistic missile program was killed because it was not possible to pacify the inhabitants of several Western states where

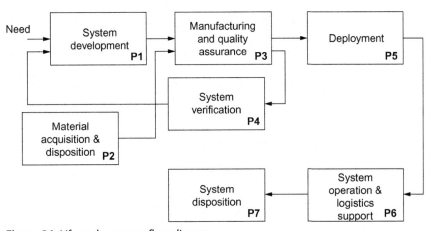

Figure 24 Life-cycle process flow diagram.

the system would be based. Every community has local smaller-scale examples of this problem in the location of the new sewerage treatment plant, dump, or prison. It is referred to as the "not-in-my-backyard" problem.

A process flow diagram often is built focusing only on process P6 and often omits the logistics processes related to maintenance and support. This is an important process and the one that will contribute most voluminously to the identification of product performance and support requirements. Expansion of P6 is what we commonly think of as the system process flow diagram. The system disposition process (P7) can also be expanded through a process diagram based on the architecture that is identified in process Pl. During process Pl, we must build this model of the system and related processes, expand each process progressively, and relate observed processes to specific things in the system architecture and processes to be used to create and support the system.

All of these processes must be defined and subjected to analysis during the requirements analysis activity and the results folded mutually consistently into product and process requirements. Decisions on tooling requirements must be coordinated with loads for the structure. Product test requirements must be coordinated with factory test equipment requirements. Quality assurance inspections must be coordinated with manufacturing process requirements. There are, of course, many, many other coordination needs between product and process requirements.

This book is focused primarily on work that occurs within Pl and P4 of Figure 24. All of the validation work will normally occur in Pl and much of the planning work for verification as well. Some of the verification work accomplished by analysis will occur in Pl, but P4 provides the principal machinery for proving that the product satisfies its requirements through testing. In order to accomplish this work, it is necessary to manufacture a product that can be tested. Some of this product may not be manufactured on a final production line but in an engineering laboratory, so the separation of development (Pl) and verification (P4) by the manufacturing (P3) process does not clearly reflect every possible development situation. In any case, where the verification process indicates a failure to satisfy requirements, it may stimulate a design change that must flow back through the process to influence the product subjected to verification testing.

3.3.2 Computer Software Derived Models

It is not possible for a functioning system to exist that is entirely computer software, because software requires a machine medium within which to function. Systems that include software will always include hardware, a computing instrument as a minimum, and most often will involve people in some way. Software is to the machine as our thoughts, ideas, and reasoning are to the gray matter making up our mind. While some people firmly believe in out-of-body experiences for people, few would accept a similar situation for software, even though a *Star Trek* episode featured software operating in outer space independently of any "hardware" aspect.

A particular business entity may be responsible for creating only the software element of a system and, to them, what they are developing could be construed as a system, but their product can never be an operating reality by itself. This is part of the difficulty in the development of software; it has no physical reality. It is no wonder then that we might turn to a graphical and symbolic expression as a means to capture its essence.

We face the same problem in software as hardware in the beginning. We tend to understand our problems first in the broadest sense. We need some way to capture our thoughts about what the software must be capable of accomplishing and to retain that information while we seek to expand upon the growing knowledge base. We have developed many techniques to accomplish this end over the period of 50–60 years during which software has become a recognized system component.

The software focus has been most often oriented toward computer processing and computer data aspects of the software. The schism between process-oriented analysis and data-oriented analysis, which was somewhat patched together in earlier analysis methods, was perhaps joined together more effectively in object-oriented analysis (OOA), about which there have been many books written. A series that is useful and readable is by Coad and Yourdon (Volumes 1 and 2, *Object Oriented Analysis* and *Object Oriented Design*, respectively) and Coad and Nicola (Volume 3, *Object Oriented Programming*). Two others are James Rumbaugh et al., *Object Oriented Modeling and Design*, and Grady Booch, *Object-Oriented Analysis and Design with Applications*. The author believes the OOA approach to have been a dead end, because all of the books on the subject encouraged that the analyst first focus on determining what the objects should be and this is in conflict with the fundamental systems engineering belief first uttered by Louis Sullivan: "form ever follows function." Thus, the reader will find no support in this book for OOA.

The several models described in subordinate paragraphs were originally independently conceived and advanced primarily for the development of computer software, but the author has found that we can combine some of these to form what was called earlier in this chapter *universal architecture description frameworks* (UADF). Functional modeling discussed earlier combined with three other UADFs provides us with four comprehensive modeling methods, any one of which an enterprise can apply to comprehensively develop the architecture and related specifications for systems and all of their product and interface entities, no matter how we will choose to implement them in terms of hardware, software, or people doing things.

3.3.2.1 Modern Structured Analysis and Process for System Architecture and Requirements Engineering

The earliest software analytical tool was flowcharting, which lays out a stream of processing steps similar to a functional flow diagram (commonly in a vertical orientation rather than horizontal, probably because of the relative ease of printing them on line printers), where the blocks are very specialized functions called computer processes. Few analysts apply flow diagramming today, having surrendered to data flow diagramming (DFD) used in modern structured analysis (MSA), the Hatley-Pirbhai extension of this technique now referred to as *process for system architecture and requirements engineering* (PSARE), *object-oriented analysis*, or *unified modeling language* (UML). Alternative techniques have been developed that focus on the data that the computer processes. The reasonable adherents of the process and data orientation schools of software analysis would today accept that both are required, and some have made efforts to bridge this gap.

All software analysis tools (and hardware-oriented ones as well) involve some kind of graphical symbols (bubbles or boxes) representing data or process entities connected by lines, generally directed ones. MSA begins with a context diagram formed by a single bubble representing the complete software entity connected to a ring of blocks that correspond to external interfaces that provide or receive data. This master bubble corresponds to the need, or ultimate function, in functional analysis, and its allocation to the thing called a system. The most traditional expression of MSA was developed principally by Yourdon, DeMarco, and Constantine. It involves expansion of the context diagram bubble into lower-tier processing bubbles that represent subprocesses, just as in functional analysis. These bubbles are connected by lines indicating data that must pass from one to the other. Store symbols are used to indicate a need to temporarily store a data element for subsequent use. These stores are also connected to bubbles by lines to show source and destination of the data. Since the directed lines represent a flow of data between computer processing entities (bubbles), the central diagram in this technique is often referred to as a data flow diagram.

In all software analysis techniques, there is a connection between the symbols used on the diagrammatic portrayal to text information that characterizes the requirements for the illustrated processes and data needs. In the traditional line-and-bubble analysis approach, referred to as data flow diagramming, one writes a process specification for each lowest-tier bubble on the complete set of diagrams and provides a line entry in a data dictionary for each line and store on all diagrams. Other diagrams are often used in the process specification to explain the need for controlling influences on the data and the needed data relationships. All of this information taken together becomes the specification for the design work that involves selection of a specific machine upon which the software will run, a language or languages that will be used, and an organization of the exposed functionality into "physical"modules that will subsequently be implemented in the selected language through programming work. A good general reference for process and data-oriented software analysis methods is Yourdon's *Modern Structured Analysis*. Tom DeMarco's *Structured Analysis and System Specification* is another excellent reference for these techniques.

Much of the early software analysis tool work focused on information batch processing because central processors, in the form of large mainframe computers, were in vogue. More recently, distributed processing on networks and software embedded in systems has played a more prominent role, revealing that some of the earlier analysis techniques were limited in their utility to expose the analyst to needed characteristics. Derek Hatley and the late Imtiaz Pirbhai offered an extension of the traditional approach in their *Strategies for Real-Time System Specification* to account for the special difficulties encountered in embedded, real-time software development. They differentiate between data flow needs and control flow needs and provide a very organized environment for allocation of exposed requirements model content to an architecture model. The specification consists of information derived from the analytical work supporting both of these models. In a subsequent book titled *Process for System Architecture and Requirements Engineering (PSARE)* the authors improved the description of their model.

PSARE in fact was devised as a system model where the DFD lines can be implemented in hardware (physical interface) or software (data) and the functionality represented by the bubbles accomplished in a digital computer or other machinery. Figure 25 provides a view of a system modeled at the highest level in a DFD with the bubbles captured within superbubbles that allocate their functionality to particular physical entities that may be

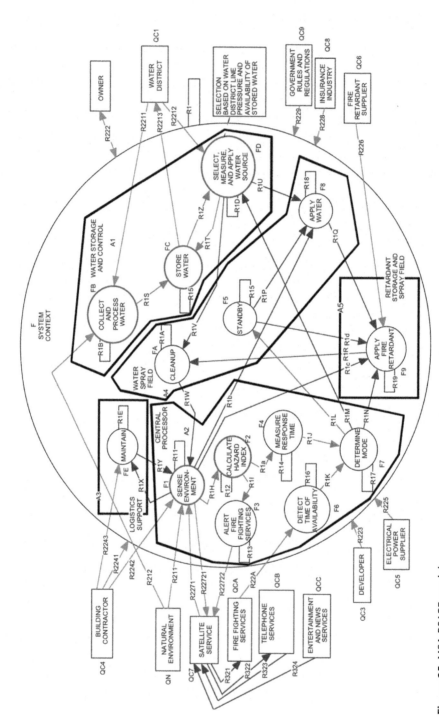

Figure 25 MSA-PSARE model.

computers or other equipment. The system depicted is intended to protect property from wind fires such as those that happen in Southern California in the fall season due to high atmospheric pressure in the interior, driving dry air at high temperature down toward the sea.

MSA and PSARE are combined into a single model by the author, forming a UADF. Neither MSA nor PSARE include modeling structures to deal effectively with specialty engineering and environmental requirements, so the author tags on those models as described later in this chapter. PSARE actually includes a requirements model and an architecture model and the control flow diagram is an important element of the overall model. MSA can, however, be used following PSARE rules, permitting implementation in hardware or software yielding a simple system modeling method without the CFD, where control functionality may not be the driving concern.

3.3.2.2 Unified Modeling Language – System Modeling Language

From this dynamic history emerged *unified modeling language* (UML), which has become the software development standard. UML did correct a serious flaw in early OOA by encouraging the application of Sullivan's notion of "form ever follows function"to software development again. UML was extended to create *system modeling language* (SysML) to apply the UML model to the development of general systems and hardware entities. Figure 26 provides a summary view of the UML approach, which can also be applied more generally to system development.

System functionality is identified in terms of use cases, expressed as actors depicted as stick figures. While not part of UML, the author encourages the use of a context diagram from MSA-PSARE to organize the use cases applied in UML. For each terminator of the context diagram we can identify some number of use cases. The actors in the use cases, which need not be humans, are said to derive benefits that are further explored by a set of dynamic models (state diagram, activity diagrams similar to functional flow diagrams, and sequence diagrams). Performance requirements derived from this dynamic analysis are allocated to product entities and interface requirements can be derived from these models. This process continues until the product entities are classes about which software code can be written so as to satisfy the requirements identified from the dynamic analysis. UML and SysML also do not include modeling structures to deal effectively with specialty engineering and environmental requirements, so the author tags on those models to UML-SysML.

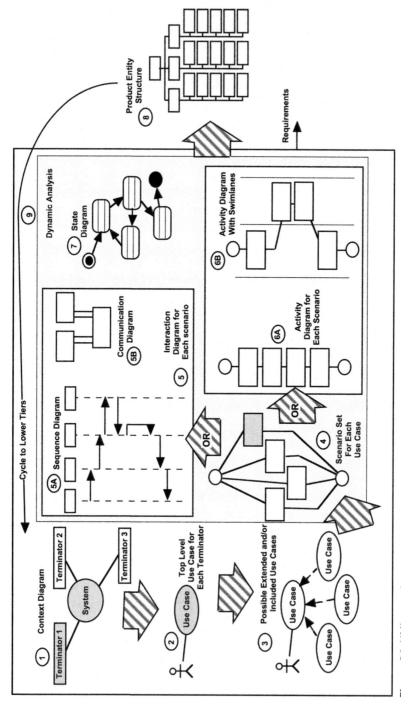

Figure 26 UML overview.

3.3.2.3 Unified Process for DoDAF MoDAF

The Department of Defense (DoD) has developed an architecture frame-work called DoDAF for use in the development of large-scale information systems needed to join multiple systems into effective systems of systems to deal with very complex threats. The United Kingdom Ministry of Defence (MoD) has adapted the model to the development of systems for the British services. DoD and MoD have cooperated in the adaptation of the union of DoDAF and MoDAF using the modeling artifacts of UML. The name of the resultant construct is named the *Unified Process for DoDAF MoDAF* (UPDM). Unfortunately, even though this modeling construct includes 52 modeling artifacts, it remains focused only on information systems and excludes modeling artifacts for identification of specialty engineering and environmental requirements. The author in the companion book *System Requirements Analysis* adds these two modeling constructs, resulting in a UADF that can be effective in both architecting systems to deal with very complex functionality and providing the modeling basis for derivation of all requirements for program specifications in the context of UML-SysML.

3.3.3 Product and Relationship Entity Identification Models

One of the two principal outputs of the modeling work discussed to this point is the performance requirements that are allocated to product entities, thus gaining insight into what those entities will be. As discussed earlier, a simple hierarchical product entity block diagram can be used as an effective model of those entities. This leaves a need for a model for the relationships between product entities in the form of interfaces. In functional analysis insight into needed interfaces comes from how the performance require-ments derived from functions are allocated to product entities. The insight that we gain from these relationships is commonly reported in the form of a schematic block diagram or n-square diagram.

All of the four UADF discussed include a means by which product enti-ties that will accomplish exposed functionality or behavior will be accom-plished. In functional analysis one must decide what entity will accomplish requirements derived from functions. MSA and PSARE employ bubbles to depict functionality and the same process can be applied there. In UML-SysML one identifies product entities in the problem space model using the swim lanes of activity diagrams and lifelines of sequence diagrams. UPDM employs UML-SysML artifacts, so this transfers to UPDM. The author would encourage that the hierarchical product entity structure be employed in all UADF.

In MSA-PSARE the bubbles are joined by functional relationship directed line segments and when we overlay the bubbles with superbubbles, we define the boundaries of the product entities, thus also exposing the physical interfaces between them. In UMl-SysML relationships between the product entities in the form of sequence diagram lifelines reveal interface relationships.

3.3.4 Design Constraints Analysis

Design constraints are boundary conditions within which the designer must remain while satisfying performance requirements. All of them can be grouped into three kinds, described in the following paragraphs. Performance requirements can be defined prior to the identification of the things to which they are ultimately allocated. Design constraints generally must be defined subsequent to the definition of the item to which they apply. Performance requirements provide the bridge between the problem and solution planes through allocation. Once we have established the product entity structure using one of the four modeling constructs just discussed, we can apply three kinds of constraints analysis to these items. In the case of each constraint category, we need a special modeling set to help us understand in some organized way what characteristics we should seek to control. Some of the modeling constructs discussed previously include interface models, but none of them include effective specialty engineering or environmental modes, so at least these two models would have to be added to form a truly comprehensive model.

3.3.4.1 Interface Requirements Analysis

Systems consist of things. These things in systems must interact in some way to achieve the desired functionality. A collection of things that do not in some way interact is a simple collection of things, not a system. An interface is a relationship between two things in a system. This relationship may be completed through many different media, such as wires, plumbing, a mechanical linkage, or a physical bolt pattern. These interfaces are also characterized by a source and a destination – that is, two terminals, each of which is associated with one thing in the system. Our challenge in developing systems is to identify the existence of interfaces and then to characterize them, each with a set of requirements mutually agreed upon by those responsible for the two terminals.

Note the unique difference between the requirements for things in the system and interfaces. The things in systems can be clearly assigned to a single person or team for development. Interfaces must have a dual responsibility where the terminals are things with different responsibilities. This complicates the development of systems because the greatest weakness is at the interfaces, and accountability for these interfaces can be avoided by one terminal party, the other, or both if the program does not apply a sound interface development management approach. The opportunities for accountability avoidance can be reduced by assignment of teams responsible for development as a function of the product entity structure rather than the departments of the functional organization. This results in perfect alignment between the product cross-organizational interfaces (interfaces with different terminal organizational responsibilities) and the development team communication patterns that must take place to develop them. Responsibility and accountability are very clear.

The reader is encouraged to refer to the author's *System Requirements Analysis* for a thorough discussion of schematic block and n-square diagramming techniques. As a result of having applied these techniques during the product entity structure synthesis of allocated functionality, the system engineer will have exposed the things about which interface requirements must be written. Once again, the purpose of our tools is to do just this, to help us understand what to write requirements about. The use of organized methods encourages completeness and avoidance of extraneous requirements that have the effect of increasing cost out of proportion to their value.

Once it has been determined what interfaces we must respect, it is necessary to determine what technology will be used to implement them, such as electrical wiring, fluid plumbing, or physical contact, for example. Finally, the resultant design in the selected media is constrained by quantified requirements statements appropriate to the technology and media. Each line on a schematic block diagram or intersection on an n-square diagram must be translated into one or more interface requirements that must be respected by the persons or teams responsible for the two terminal elements. The development requirements for the two terminal items may be very close to identical, such as a specified data rate, degree of precision, or wire size. The product requirements, however, will often have an opposite nature to them, such as male and female connectors, bolt hole or captive bolt and threaded bolt hole, or transmitter and receiver.

3.3.4.2 Specialty Engineering Requirements Analysis

The evolution of the systems approach to development of systems to solve complex problems has its roots in the specialization of the engineering field into a wide range of very specialized disciplines for the very good reasons noted earlier. Our challenge in system engineering is to weld these many specialists together into the equivalent of one all-knowing mind and apply that knowledge base effectively to the definition of appropriate requirements, followed by development of responsive and compliant designs and assessment of those designs for compliance with the requirements as part of the verification activity.

There are many recognized specialized disciplines, including reliability, maintainability, logistics engineering, availability, supportability, survivability and vulnerability, guidance analysis, producibility, system safety, human engineering, system security, aerodynamics, stress, structural dynamics, thermodynamics, and transportability. For any specific development activity, some or all of these disciplines will be needed to supplement the knowledge pool provided by the more general design staff.

Specialty engineers apply two general methods in their requirements analysis efforts. Some of these disciplines use mathematical models of the system, as in reliability and maintainability models of failure rates and remove-and-replace or total repair time. The values in these system-level models are extracted from the model into item specifications. Commonly, these models are built in three layers. First, the system value is allocated to progressively lower levels to establish design goals. Next, the specialty engineers assess the design against the allocations and establish predictions. Finally, the specialists establish actual values based on testing results and customer field use of the product.

Another technique applied is an appeal to authority in the form of customer-defined standards and specifications. A requirement using this technique will typically call for a particular parameter to be in accordance with the standard. One of these standards may include a hundred requirements, and they all flow into the program specification through reference to the document unless it is tailored. Specialty engineers must, therefore, be thoroughly knowledgeable about the content of these standards; familiar with their company's product line, development processes, and customer application of that product; and knowledgeable about the basis for tailoring standards for equivalence to the company processes and preferred design techniques.

3.3.4.3 Environmental Requirements Analysis
One of the most fundamental questions in system development involves the system boundary. We must be able to unequivocally determine whether any particular item is in the system or not in the system. If it is not in the system, it is in the system environment. If an item is in the system environment, it is

either important to the system or not. If it is not, we may disregard it in an effort to simplify the system development. If it is important to the system, we must define the relationship to the system as an environmental influence. We may categorize all system environmental influences into the five following classes:

1. Natural Environment: Space, time, and the natural elements such as atmospheric pressure, temperature, and so forth. This environment is, of course, a function of the locale and can be very different from that with which we are familiar in our immediate surroundings on Earth, as in the case of Mars or the Moon.

2. Hostile Systems Environment: Systems under the control of others that are operated specifically to counter, degrade, or destroy the system under consideration.

3. Noncooperative Environment: Systems that are not operated for the purpose of degrading the system under consideration but have that effect unintentionally.

4. Cooperative Systems Environment: Systems not part of the system under consideration that interact in some planned way. Generally, these influences are actually addressed as interfaces between the systems rather than environmental conditions, because there is a person from the other system with whom we may cooperate to control the influences.

5. Self-Induced Environment: Composed of influences that would not exist but for the presence of the system. These influences are commonly initiated by energy sources within the system that interact with the natural environment to produce new environmental effects.

As noted here, cooperative environmental influences can be more successfully treated as system interfaces. Hostile and noncooperative influences can be characterized through the identification of threats to system success and the results joined with the natural environmental effects. The induced environment is best understood through seeking out system energy sources and determining if those sources will interact with the natural environment in ways that could be detrimental to the system.

The natural environment is defined in standards for every conceivable parameter for Earth, space, and some other heavenly bodies. The challenge to the system engineer is to isolate those parameters that are important and those that are not, and then to select parameter ranges that are reasonable for those parameters that will have an impact on our system under development. The union of the results of all of these analyses forms the system environmental requirements. It is not adequate to stop at this point in the analysis, however.

Systems are composed of many things that we can arrange in a family hierarchy. Items in this hierarchy that are physically integrated in at least some portions of their system operational use, such as an aircraft in an aircraft system or a tank in a ground combat system, can be referred to as end items. We will find that these end items in operational use will have to be used in one or more environments influenced by the system environment but, in some cases, modified by elements of the system. For example, an aircraft will have to be used on the ground, in the air through a wide range of speed and altitude, and in hangars. These are different environments that we can characterize as subsets of the system environment definition. The best way to do this is to first define the system process in terms of some finite number of physical analogs of system operation. We may then map the system end items to these process steps at some level of indenture. Next, we must determine the natural environmental subsets influencing each process step. This forms a set of environmental vectors in three-space. In those cases where a particular end item is used in more than one process, we will have to apply some rule for determination of the aggregate effect of the environments in those different process steps. The rule may be worst case or some other one. This technique is called *environmental use profiling*.

The final step in environmental requirements analysis involves definition of component environmental requirements. These components are installed in end items. The end items can be partitioned into zones of common environmental parameters as a function of the end item structure and energy sources that will change natural environmental influences. We define the zone environments and then map the components into those zones. In the process, we may find that we have to provide an environmental control system in one or more of the zones to reduce the design difficulty of some components. We may also conclude, given that we must have at least one such system, that we should relocate one or more components into the space thus controlled. So, the environmental requirements for a component are predetermined by the zone within which the component is located, and we may derive its environmental requirements by copying (cloning) the zone requirements. Component environmental requirements may have to be adjusted for shipping and other noninstalled situations.

In summary, this set of three tools (standards, environmental use profiling, and end item zoning) may be used to fully characterize the environmental requirements for all system items from the system down through its components. In all cases, we must be careful to phrase these requirements in terms that recognize our inability to control the ultimate natural environment.

The preceding discussion focuses on the effects of the environment on the system. We must also consider the effects of our system on the natural environment. This is most often referred to as environmental impact analysis. It has not always been so, but today we must be sensitive to a bidirectional nature in our environmental specification. Our efforts in the distant past were very small compared to the tremendous forces of nature. Today, the scope of our access, control, and application of energy and toxic substances is substantial, and potential damage to local and regional natural environments is a real concern. Environmental impact requirements are commonly defined in laws and regulations. We can determine in what ways the environment will be exposed to system-generated stress by evaluating all system energy sources, toxic substances used, and any exhaust products. The author prefers to classify this topic under the banner of environmental requirements analysis, but it could be thought of as a specific specialty engineering discipline.

These effects must be considered both for the life of the system during its use and, perhaps more importantly, for its disposition following its useful life. There can be no better example of how difficult and dangerous this can be than the case of the nuclear weapons cleanup process seriously begun in the 1990s after the end of the nuclear confrontation between the USSR and the United States, lasting 40 years. Throughout the confrontation, both sides were so concerned with survival and the urgency of development that little thought was given to the problems that they might be causing for the future. By the time the problem could no longer be ignored, it was so substantial that a solution was more difficult than the complex scientific and technical problems that had to be solved to create it. One can be very unpopular expressing interest in system disposition during the heady system development times, especially with one's proposal or program manager. But, customers will increasingly be concerned with this matter as the problems we attempt to solve become increasingly more general and their scope more pervasive.

Computer software environmental requirements are limited. The natural environment impinges on the computer hardware within which the software functions, but does not in any direct way influence the software itself in the absence of hardware deficiencies that cause computer circuits to fail due to environmental stresses. The context diagram used to expose the boundary between the software and its environment is the best place to start in determining the appropriateness of any environmental influences directly on the software. One such requirement category of interest commonly is the user environment, describing how the user will interact with the software. These influences could alternatively be covered under the heading of interface requirements. One must first determine whether the operator is within or outside the system, however, and this is no small question.

The principal element in the software environment is in the form of cooperative systems exposed through interface analysis. What other software and hardware systems will supply information or use information created by the system in question? The noncooperative system component may exist in very complex software systems, but it is hard to characterize in terms of an example. A very common and increasingly important environmental factor is the hostile environment. Noncooperative and hostile systems seldom have a direct effect on the operation of the software, but can have indirect effects through invasion and corruption of the computer through network access.

3.4 TRANSITION POINT

When the item performance specification is complete for an item, the responsible team can be reoriented to move to the next phase of development work for that item. Completion of the item performance specification means that it has been reviewed by management and approved for release. The master for the specification is locked up by configuration management, preventing any changes without a formal specification change and review and approval of that change. The specification should include both the product requirements (Section 3) and the verification requirements (Section 4). The specification should include within it a verification traceability matrix, best located in Section 6, the author believes, with the other traceability data.

As item concept and preliminary design work comes to a close, the system engineer responsible for program verification on the program integration team can begin working with the item team member made responsible for item qualification verification to identify verification tasks, assign task principal engineers, develop verification task budgets and schedules, and educate those principal engineers in their work to come.

3.5 AN IMPORTANT MANAGEMENT RESPONSIBILITY

Enterprises that develop new systems are often poorly prepared to accomplish the systems development requirements work, and where this work is done poorly the poor results propagate through the remainder of the program work, resulting in a less than satisfied customer. Management has a means to correct this kind of problem, but in all too many firms does not take the necessary actions to make progress. An important part of the corrective action is the selection of one of the four models briefly discussed in this chapter, training the staff to use it well, and insistence on its application on all programs to gain the substantial training benefit from repetition. The resulting improvement in the content of program specifications will ripple through synthesis and verification work on every program, yielding improved customer attitudes about the systems delivered to them.

It is often claimed that the very complex software development programs experienced today cannot follow the rigid pattern suggested in this book because it is not possible to know all of the requirements early in a program. If this is a correct position, then on such programs it may be necessary to apply the agile development model briefly discussed in Chapter 1. This model does not, however, reduce the utility of requirements to zero. On such a program one must capture a definite set of requirements as best as one can in each program cycle and, if necessary, in a subsequent cycle make changes to that set upon which the work of the next cycle will be based.

CHAPTER 4

Specification Section 4 Preparation

4.1 CHAPTER OVERVIEW

Product requirements appearing in Section 3 of a specification define essential characteristics that the design of a product must possess, and the design team assigned to that task is responsible for creating a design that will comply with those requirements. In order to ensure that these design teams satisfy their responsibility, we derive verification requirements from the product requirements, including them in a quality assurance or verification section of your specification (Section 4 of a military specification). For every performance specification requirement included, there should be one or more verification requirements that tell how it will be determined whether or not the design solution complies with that requirement. Verification requirements define the essential characteristics for the design of verification tasks. All verification requirements are assigned to a verification task of one of the five classes. The aggregate of all verification tasks identified based on the content of one specification must fully examine the design of the product driven by the aggregate of product requirements in that specification.

Paragraphs 4.1 through 4.3 of Section 4 for a performance specification, as the reader can see from Figure 16, are all introductory and programmatic in nature. Paragraph 4.4 contains all of the detailed verification requirements derived from Section 3 of the specification. No special model is put forward for deriving the verification requirements. This work requires good engineering knowledge applicable to the attribute called out in the requirement. Requirements dealing with fluids, electrical power or signals, or mechanical motion or action of some kind will require either stimulation or sensory capabilities, depending on whether the requirement relates to an input or output, respectively. The verification requirements need only provide a measurement based on a particular situation and it will be the responsibility of the verification task principal engineer to which the requirement is assigned to determine how in the task plan and procedure to present the measurement to the person who will run the task.

System Verification
http://dx.doi.org/10.1016/B978-0-12-804221-2.00004-8

The reader will note that the author's template in Figure 16 does not provide guidance on the order of listing or sequencing of the verification requirements or any guidance on how to partition all of the verification requirements into task collections. This is the plane that separates requirements analysis work from verification planning work. This is where a system engineer skilled in verification planning will take over when the specification has been released. The specification should include a verification traceability matrix (such as that shown as Table 6-5 in Figure 19 of this book) listing all of the product requirements in the left-hand column in paragraph number order, title, and the corresponding verification requirement paragraph number followed by a definition of verification methods. The latter can be provided for by a column for each of the accepted methods, which may include test, analysis, examination, demonstration, special, and none required. An *X* is placed in the appropriate columns for each product requirement. Alternatively, the method name can be spelled out in a method column. It is possible that a single requirement could require multiple methods, though this is uncommon. A column should also be included for the level at which the verification compliance action shall be determined, with the normal entry being *item* but other possible entries being *parent* or *child*. There should be one or more verification requirements defined for each *X* (except the none case) in the matrix.

The reason that we must identify the level of system hierarchy at which the requirement will be verified is that some requirements in the performance specification for product entity A12 could be promoted to a verification task associated with product entity A1 or demoted to a verification task associated with product entity A122. For example, if the requirement for an aircraft altitude control unit requires that it maintain aircraft barometric altitude to within 100 feet, we could require at the system level that a flight test demonstrate this capability with actual altitude measured by a separate standard and that it not deviate by more than 100 feet from the altitude commanded. At the avionics system level, this may be verified through simulation by including the actual altimeter in an avionics system test bench with measured altitude error under a wide range of conditions and situations. At the black-box level, this may be stated in terms of a test that measures an electrical output signal against a predicted value for given situations.

The requirements for the verification tasks accomplished to determine the extent to which the design solution satisfies the product requirements must be captured in some form and used as the basis for those tasks. In specifications following the military format, Section 4, titled Quality Assurance in MIL-STD-490 or Verification in MIL-STD-961E, has been used to do this, but in many organizations this section is only very general in nature, with the real verification requirements included in a separate verification requirements document or an integrated test plan or procedure. Commonly, this results in coverage of only the test and demonstration requirements, with analysis and examination requirements being difficult to find and to manage. The best solution to this problem is to include all of the verification requirements in the corresponding specification, to develop them in concert with the performance requirements and constraints, and to use them as the basis for test, analysis, examination, demonstration, and special verification planning work that is made traceable to those requirements.

Ideally, a system engineer would build the verification traceability matrix for each specification, stating method and level for each requirement appearing in Section 3. There will be many opportunities for the team members of the responsible team to discuss methods and levels for the product requirements over the period the preliminary design matures, so the system engineer should use these as opportunities for integration and optimization in search of the best way of accomplishing verification. Ideally, every verification requirement included in Section 4 of a specification would be written by the engineer responsible for writing the corresponding product requirement included in Section 3.

In the next chapter we will expand the simple verification traceability matrix discussed here into a program-wide requirements compliance matrix used in the management of the whole verification process. This matrix will tie together the information provided for each specification with the verification tasks planned to create evidence of the degree of compliance and the evidence resulting from those tasks. This matrix will be one of the most important documents in the life of the program. Very important work is accomplished by the system engineers working on this table and the engineers providing information inputs to the table via the content of the verification traceability matrices for the product items for which they are developing specifications. An enterprise should formulate guiding criteria for these engineers so as to avoid excessive cost as well as holes in the verification story. While verification planning strictly focused on the isolated content of each specification will guarantee complete program verification coverage, it may introduce unnecessary cost into the process where requirements are verified at multiple product levels. These are often very difficult choices to make, but the alternatives should be evaluated.

4.2 VERIFICATION CLASSES

There are four classes of verification. Each class correlates with a kind of specification. Figure 27 illustrates how the two sets composed of specifications and verifications classes align. The system specification is only prepared in a performance specification format with no detail specification. At the item and interface level frequently both parts are prepared; parts, materials, and processes (PMP) specifications may include both performance and detail content.

All of the types of specifications can be applied to hardware or software – yes, even PMP. It is possible for a software system specification to be prepared but we have to recognize that software can never run without a hardware component (except for that episode on the *Star Trek* television show where Dr. Spock detected software running in space observed as unexplained flashing lights).

4.2.1 Parts, Materials, and Processes Class

First, we should verify that all parts, materials, and processes identified on program engineering drawings and lists comply with specifications for those entities. Often, specifications for these entities will be prepared by suppliers

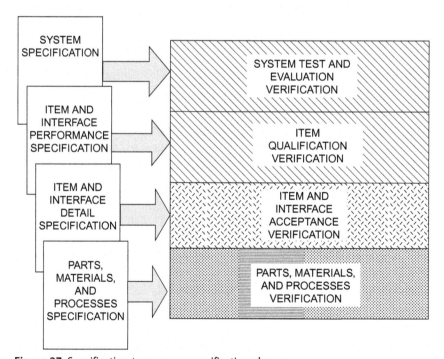

Figure 27 Specification types versus verification classes.

for these products, but many are developed by engineering societies and standards agencies. Chapter 7 focuses on this class of verification.

A program should set up a standard parts, materials, and processes activity early before any preliminary design work has been accomplished. This activity will provide design engineers with a list of all approved parts, materials, and processes they may call on drawings. Before an item can be listed, it should be verified that it will be suitable for use on the program. This is essentially a verification process most often accomplished by examining the capabilities of the item relative to program requirements, but in other cases testing may have to be accomplished. If a design engineer concludes there is no adequate part, material, or process that will satisfy the need on the current approved list, then the engineer must request that a new item be added to the list, and in order to be approved it will have to be determined that the item complies with program requirements and that there is no other item on the list that would be adequate. The desire is to minimize the number of different items on the program list.

Different enterprises organize their PMP departments differently. You might find parts under electrical engineering, materials under mechanical engineering, and processes elsewhere. It is possible that all three could be in one PMP department or any arrangement between these extremes. In any case, it should be clear that there is a strong correlation between a standard PMP list and verification. Sadly, few enterprises recognize the parts, materials, and processes class as appropriate for computer software.

4.2.2 Item and Interface Qualification Class

The parts and materials selected will, of course, appear in product item and interface design solutions intended to comply with the requirements in the performance, item development, or Part I specification for the item or interface. Some engineers refer to this as verifying the "design-to requirements." This process is more formally referred to as qualification of the item for the application. The associated verification work is commonly only accomplished once in the life of the item, using a dedicated engineering article produced just for qualification. The program manufacturing capability will commonly not have matured sufficiently to produce the articles required in the qualification process, but these low-rate production units, often built in an engineering laboratory, should be as representative of the production units as possible to ensure that the verification results can be extrapolated to full production units.

A contract will often call for a functional configuration audit (FCA) for each product entity requiring qualification. This audit will require the contractor to present for customer audit such documentation as the customer wishes to see in order to reach a sound conclusion about the degree to which the product complies with the requirements. It should be noted that the article subjected to qualification may not survive the qualification process in that testing is often applied after it has been determined that requirements have been complied with, to determine how much margin there is in the design. Positive margins are generally good, but excessive margin can contribute to unnecessary cost over the run of the program. Chapter 8 focuses on item and interface qualification verification.

4.2.3 Item and Interface Acceptance Class

The third verification class involves acceptance of each product article as it becomes available from manufacturing or other sources to determine if it is acceptable for sale or delivery. In this case, we seek to verify that the individual article complies with "build-to requirements" captured in an item product, Part II, or detail specification. These specifications were renamed *detail specifications* in MIL-STD-961D, Appendix A, and that name was carried on in the E revision. This activity could also be driven by engineering drawings as well as quality and manufacturing planning documentation without reference to detail specifications. This process is referred to as product acceptance and is either accomplished on each product article delivered or within the guidelines of some sampling regime.

Some organizations insert a first article inspection between these two classes that may call for special inspections that will not be required in recurring conformance inspection of each production article. In this case, the acceptance applied to all of the other articles coming through the production process is referred to as a *conformance inspection*. The author has collapsed these two into a general acceptance class in this book, but the reader may wish to separate these and differently characterize them. An example of a case where this might be useful is where the first article is being used as a trailblazer to check out the process as well as manufacture or process the product. Also, it is the first article inspection results that will drive the physical configuration audit.

Some organizations conduct a combined first article and qualification verification, but there are some real problems with this approach that encouraged DoD to move from this approach to the two-step approach that has been used for many years. During a proper qualification process, the product is stimulated by the full range of stresses its specification calls for, to prove whether the design is adequate. Environmental requirements extremes like vibration, shock, thermal, and pressure especially tend to cause an item subjected to qualification to be deprived of some of its life, if it operates at all at qualification completion. Thus, an item subjected to qualification cannot be sold to the customer. Later we will discuss some useful ways to use such units, but a normal life in the hands of the customer is not one of them. It therefore makes sense to accomplish item verification in two steps. The first step proves the design satisfies its development requirements, while the second step proves that the manufacturing process will produce a product that satisfies the engineering and manufacturing requirements. Chapter 10 focuses on item and interface acceptance verification.

4.2.4 System Development Test and Evaluation Class

The system specification is similar in content to an item performance specification so there are a lot of similarities between item qualification and system test and evaluation. System testing is commonly partitioned into two subclasses: (1) development test and evaluation (DT&E), which is conducted by the contractor to determine the extent to which the system performance complies with the content of the system specification; and (2) operational test and evaluation (OT&E), conducted by the user of the system to determine the extent to which the product can satisfy user mission needs. It has happened before that a customer acquisition agent will let a contract, including a system specification, the content of which fails to fully comply with the content of the user requirements document, because there simply is not enough money in the budget to fully comply with user requirements in terms of speed, payload capacity, endurance, and other factors. It has also happened that a system will pass DT&E yet fail OT&E for this very reason. The right time to prevent this problem from occurring on a program is at the system requirements review (SRR), which is scheduled to take place very early in a program in order to reach a clear understanding between contractor and customer about the content of the system specification and to reach agreement that it is the content of the system specification that contains the formal list of requirements for the system under contract.

Generally, DT&E is not supposed to commence until all qualification verification work is complete in the interest of safety, especially if people will have to participate in or on the system end items during system testing, as in flying airplanes for example. DT&E may also require the use of special test ranges that provide a large tract of land, sea, and air space within which failures will not result in hazards to personnel or property not involved in the program under test. Chapter 9 covers system test and evaluation.

4.3 ITEMS SUBJECT TO QUALIFICATION AND ACCEPTANCE

Qualification and acceptance verification is commonly accomplished on items variously referred to as *configuration* or *end items*. These are items through which a whole development program is managed, items that require new development and have a physical or organizational integrity, and items that will be assigned to cross-functional teams responsible for development of the design or procurement. They are designated in early development planning and major program design reviews and audits are organized around them. Ideally, every branch in the system product entity structure can be traced up through at least one of these items such that verification work accomplished on them will ensure that the complete system has been subjected to verification. Very complex systems may include more than one level of these items.

It is very important to specifically list the items that will be subjected to formal verification of requirements. It is not, in every situation, necessary to verify everything at every level, but it is possible to make outrageous mistakes when we do not consciously prepare this list and publish it, exposing it to critical review. Unfortunately, these mistakes are not always immediately obvious and are only discovered when the product enters service. Verification of the requirements for components does not necessarily guarantee the next higher-level item will satisfy its requirements. The Hubble Space Telescope will forever offer an example of this failure. In computer software, most people would be rightly reluctant to omit testing of any entity in the hierarchy. At the same time, in some cases it should not be necessary to verify at all levels. It requires careful analysis and good engineering judgment, built through experience, to make these calls.

In the example offered earlier of the altitude hold requirement, the black box might be perfectly capable of quickly stabilizing at the ordered altitude plus or minus 100 feet, but in flight tests of the complete aircraft it may be found that unanticipated aerodynamic effects cause an altitude oscillation with an amplitude of greater than 200 feet and possibly even biased off from the commanded altitude. Past product experience can cloud our judgment. If, for example, our product had in the past always been used in altitude hold only over ground, we might discover in flight test or early user employment that the error band expanded when used over water with a sea state above or below some particular value and in particular when maneuvering in a turn.

While this book is focused on item-level and system-level verification, it should be pointed out that verification stretches all the way to the bottom of the development food chain. Parts will be verified against the claims of their parts specifications, and rolls of stainless steel will have had to be inspected for compliance with their material specification. The intent is that items will have been manufactured from qualified parts and materials and that those items will prove out when subjected to item qualification. These items when assembled into end items can be shown to function satisfactorily relative to end item specifications, and the whole system can be shown to satisfy the content of the system specification. It is possible to simply build a system and attempt to make it work as a system, but this is clearly a very risky approach compared to the careful bottom–up verification process suggested in this book. It is simply risk averse to seek to identify a design problem at the lowest level possible. It is often possible to identify problems in informal or formal design reviews, but other problems may require operation of the article and organized inspection of performance relative to its intended capability, that is, qualification.

4.4 VERIFICATION DIRECTIONALITY

Things in a system are most often verified from the bottom up where, as each item becomes available, we subject it to a planned verification sequence. The results of the verification work for all of the items at that level and in that product entity branch are used as a basis for verifying the parent item, and this pattern is followed up to the system level. This sequence is especially preferred where there is greater danger in the operation of the system the higher in the product structure one proceeds. It is a much more serious situation when a test aircraft crashes than when an altitude control box fails on a workbench. The top–down approach is not often applied in hardware but can be used in software in combination with a top–down build cycle. Initially, we create the code corresponding to the system level, complete with the logic for calls to lower-tier entities that can be simulated in system-level tests. As the lower tier entities become available, they can be installed in place of the simulating software and testing continued.

4.5 PRODUCT VERIFICATION LAYERING

A complex product will consist of many items organized into a deep family tree. When following a bottom–up directionality, it may be necessary to verify component-level items, followed by verification applied to subsystems formed of components, followed by verification of systems composed of subsystems within end items. Generally, the qualification process will subject the product to stresses that preclude its being used reliably in an operational sense and may argue against use of components in subsequent parent item qualification testing. The part of the hardware qualification testing that has this effect is primarily environmental requirements verification.

Where we apply bottom-up testing with component-level environmental testing, we may have to produce more than one article for qualification testing. After component qualification testing, these units may have to be disposed of or used in continuing functional testing in an integration lab or maintenance demonstrations. The next article will have to be brought into the subsystem testing. If that level must also be subjected to environmental testing, yet another set of components may be required for system-level qualification. In this case, the system is a collection of cooperating subsystems to be installed in an end item rather than the complete system. It might be possible to avoid multiple qualification units if the environmental testing can be delayed until after functionally oriented system integration testing, but generally we wish to be assured that all of the lower tier entities have been environmentally qualified before beginning system testing, particularly where safety issues may be involved. We will return to this problem under system verification.

Some of the verification planning will have to be considered during a proposal effort before the program has had an opportunity to clearly think through precisely what the system shall consist of in terms of end items and components. In order to cost the system, we will have to know how many of each kind of item to be qualified we will have to build. Estimating personnel will need some help from system engineering people on the proposal to assemble the best possible material cost estimate. A preliminary product entity structure should be built using the best information available at the time, and design verification engineers with knowledge of these entities polled to determine an appropriate verification approach to help quantify the number of items needed.

4.6 FORMING VERIFICATION STRINGS

The requirements in Section 3 of a specification following the template offered in Chapter 2 apply to the design of the product entity covered by the specification. The verification requirements in Section 4 apply to the process through which the degree of product compliance with the content of Section 3 will be determined. So, verification requirements apply to the preparation of the verification task plans and procedures through which the degree of product compliance will be determined.

Each product requirement in Section 3 should have one or more verification requirements prepared and included in Section 4 and the specification should include a verification traceability matrix linking Section 3 and 4 content, plus stating the method by which verification will be accomplished and the level at which it shall be accomplished. This matrix is the separation plane between requirements analysis work and verification work on a program. We will use the union of the content of all of the verification traceability matrices, one from each specification, to drive the verification planning work covered in the chapters dealing with item and interface qualification, system test and evaluation, and item and interface acceptance.

Figure 28 offers a fragment from a verification traceability matrix from a performance specification for an integrated guidance set. The matrix coordinates the requirements (in Section 3) with verification methods to be applied, a level of accomplishment intended, and a reference to the paragraph in Section 4 containing the corresponding verification requirement. The requirement identification (RID) columns for product and verification requirements are needed in requirement database systems to enable an effective traceability capability. Such systems commonly automatically uniquely assign these codes to each requirement and once assigned are never changed. In the case of traceability using paragraph numbers results in changes to these links whenever paragraph numbers are changed. In this case, a six-place base 60 system is being employed, each character of which is selected from the ten numerals of the Arabic number system, uppercase English alphabet characters less O, and lowercase English alphabet characters less l. The exclusions are to avoid human confusion between numbers and letters when visually observing one of these RID and to arrive at a convenient number of character possibilities (60). While the RID appears to employ a six-place system, the leading R is a modeling ID for a requirement that is common to all RID, so we are actually using a five-place system that will permit unique identification of 777,600,000 requirements (60^5), many more than would be needed in all of the specifications for any system. Let us say that an average

SECTION 3 PARAGRAPH	PRID	SECTION 4 PARAGRAPH	VRID	METHOD	LEVEL
3.2.12.2	R346TG	4.4.23	RpU73J	TEST	ITEM
3.3.5.1	RfYu37	4.4.26	R246Tg	EXAMINATION	ITEM
3.3.5.1	RhLqU7	4.4.13	R5U765	ANALYSIS	ITEM

Figure 28 Sample verification traceability matrix fragment.

specification includes 400 paragraphs in Sections 3 and 4, and the average program requires 200 specifications. A program will then have to deal with 80,000 paragraphs in its complete specification library. The reality is that many of these paragraphs will not require RID assigned, but let us assume that our policy is to assign RID for all paragraphs. Clearly, a five-place base 60 system will be more than adequate to uniquely identify every requirement, in that we will be able to handle 777,600,000 unique requirements on each program. We could actually back off to a four-place system, providing for unique identification of nearly 13 million requirements. Paragraph title and text have been omitted from Figure 28 in the interest of space on the book page.

The table distinguishes between RID assigned to product requirements appearing in Section 3 and verification requirements appearing in Section 4 by preceding the RID term with an R or a V, respectively. In the computer application where this data is stored and manipulated, it is not necessary to include a separate field for each kind of RID. They can both be simply identified in an RID field, and in fact could be so designated in Figure 28, but the author will distinguish between them in this book external to the computer application for added clarity.

Some engineers insist on assigning the same paragraph numbers in Section 4 that are used in Section 3, with the exception of the leading 4 being added. While this did make it easy to understand the correlation between them in typewriter technology days, it places an added and unnecessary burden on those who must maintain the document. Specification paragraph numbering will change from time to time, and it will ripple through Section 4 when this approach is used. Granted, much of this can be avoided by retaining the paragraph numbers of deleted requirements with a note "Deleted," but this is not a total solution. As the reader can see from Figure 28, this matrix provides clear traceability between the two sections using product and verification RIDs, so it matters not how you number Section 4 paragraphs so long as the numbers are unique and follow the prescribed template. Because this table provides traceability between Sections 3 and 4 of the specification, it can be included in Section 6 with all of the other traceability data. This is the author's preference, though it may be more common to place this table in Section 4.

The author will use the term *verification string* for each line item in a traceability matrix. As the reader can see, each of these strings will consist of a correlated combination of an item product requirement (paragraph number, complete paragraph, or parsed fragment), a verification method, a level at which the verification action shall take place (item, parent, or child) and its corresponding verification requirement (paragraph number, complete paragraph, or parsed fragment). There are three verification strings in Figure 28. The table could include a way of coordinating with paragraph fragments, recognizing the possibility that the specification may be written with more than one requirement in a paragraph. The rationale for a possible need for parsing paragraphs into two or more unique requirements is that all specifications will not be perfectly prepared with only one requirement in each paragraph. Even if your organization is staffed with perfect analysts, your customers and your suppliers may foist upon you imperfect specifications containing page-long paragraphs stuffed with multiple requirements. The fragmenting scheme would probably require that the verification matrix include the text of each of these paragraph fragments.

Different database systems will use different ways to form the RID character strings. Such a database system can uniquely refer to the verification strings using a concatenated field pair of Section 3 and 4 IDs as in the example PRID-VRID = R346TG-RpRU73. The hyphen is not really necessary but is used in this example for clarity in reader understanding. It is possible that the performance of the database system might be improved through the use of a separate string number field, but this gets complicated for us humans.

It is wise to refer to the requirements with a paragraph number in the matrix even though they may change over time, because the humans on the program will be able to relate more readily to paragraph numbers than requirement IDs. However, the program should be using a requirements database within which all of the specification and verification documentation content is captured. This database should automatically assign a unique RID to each product entity requirement and RID to each verification requirement. Inside of the database, it is through these RIDs that traceability is maintained no matter what changes are made in the paragraph numbers. Ideally, the database would also coordinate the possible several fragments of a paragraph with the RIDs. For example, in Figure 28 both VRIDs R246Tg and R5U765 map to specification Paragraph 3.3.5.1. This could be a case of multiple requirements in one paragraph (not a good idea, but a common reality), but alternatively an example of one requirement requiring more than one verification method, each of which would be mapped to a different verification task.

All of the RIDs assigned on a program set of specifications should be unique and may be assigned randomly by the requirements database as the requirements are added to the database. There is one case where the RID selection will be forced, however. When an item and its specification are imported from one program to another, it would be desirable to retain the same RID assignments in that specification across whichever programs it is used on, because the RID was included in the verification traceability matrix included within the specification. This can be respected if the database system, when it is in the process of assigning the RID for all of the requirements imported with a specification created on another program, can check the existing RID previously entered in the database and make any adjustments needed to permit respect for the existing RID assignments in the imported specification.

Some organizations prefer the term *verification event* to *verification string*. The author will use the term *verification string* rather than *event* because he concluded that the word *event* should be reserved for its normal planning data context to represent an instant in time corresponding to some important condition. Also, the word *string* has an immediate appeal in a physical sense because we will attach these strings of characters to verification tasks and thus build the requirements for the corresponding plan and procedure.

We have labored over this notion of verification strings because it is an extremely important point. This whole book focuses on the mapping of verification strings to verification tasks as the means to design an affordable verification process. We will recognize five kinds of tasks, each coordinated with one of the five verification methods, to be discussed next. A particular test task may have 31 test strings mapped to it. Every verification string for every specification in the system will have to be mapped to one or more verification tasks. Then each of those tasks will be assigned to a particular responsible engineer who must plan the task, build a procedure, acquire the resources with which to accomplish the task, accomplish the task on schedule within the cost and schedule allowed, and report the results.

Figure 29 illustrates an example of a fairly complex paragraph in a specification Section 3 coordinated with the sample content of Figure 28. The paragraph consists of three fragments, each of which must be verified through an application of a different combination of verification methods. Each combination of a fragment, a method, level of accomplishment, and the corresponding verification requirement in Section 4 corresponds to a unique verification string shown on the line connecting Section 3 data to Section 4 data by a concatenated PRID-VRID string. In the case where a paragraph was properly written and contains only a single fragment, and that requirement will be verified by a single method, this identification approach will collapse into a single verification string assigned for the paragraph. Throughout the discussion of verification in this book, the reader will find that the author has tried to employ the most complicated possible situations that will cover the simple cases we may more often come in contact with.

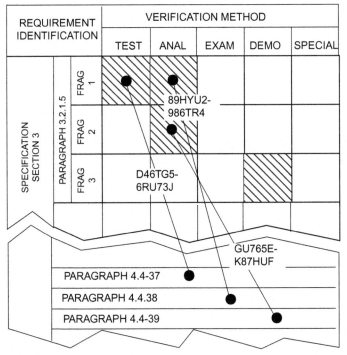

Figure 29 Verification strings.

4.7 VERIFICATION METHODS

Figure 29 identifies the methods to be used in verifying the listed requirements. The five principal methods employed are: (1) analysis, (2) demonstration, (3) examination, (4) test, and (5) special, though to be complete we should also add "none." The examination method was referred to as "inspection" in MIL-STD-490A, but this often resulted in confusion because all verification methods were also referred to under the general heading of quality inspections, in the sense that quality assurance engineers use this term. MIL-STD-961E uses the term *inspection* as a generic method, of which the other five are examples, and uses the word *examination* for one of the specific methods in place of *inspection*. We add "none" to cover the cases where a paragraph includes no requirements, as in header-only kinds of paragraphs like "3. Requirements."

4.7.1 Analysis

Analysis is an element of verification or inspection that utilizes established technical or mathematical models or simulations, algorithms, charts, graphs, circuit diagrams, or other scientific principles and procedures to provide evidence that stated requirements were met. Product item features are studied to determine if they comply with required characteristics.

4.7.2 Demonstration

Demonstration is an element of verification or inspection that generally denotes the actual operation, adjustment, or reconfiguration of items to provide evidence that the designed functions were accomplished under specific scenarios. The items may be instrumented and qualitative limits of performance monitored. The product item to be verified is operated in some fashion (operation, adjustment, or reconfiguration) so as to perform its designated functions under specific scenarios. Observations made by engineers are compared with predetermined responses based on item requirements. The intent is to step an item through a predetermined process and observe that it satisfies required operating characteristics. The items may be instrumented and quantitative limits of performance monitored and recorded.

4.7.3 Examination

Examination is an element of verification or inspection consisting of investigation, without the use of special laboratory appliances or procedures, of items to determine conformance to those specified requirements which can be determined by investigations. Examination is generally nondestructive and typically includes the use of sight, hearing, smell, touch, and taste; simple physical manipulation; mechanical and electrical gauging and measurement; and other forms of investigation. A test engineer makes observations in a static situation. The observations are normally of a direct visual nature unsupported by anything other than simple instruments like clocks, rulers, and other devices that are easily monitored by the examining engineer. The examination may include review of descriptive documentation and comparison of item features and characteristics with predetermined standards to determine conformance to requirements without the use of special laboratory equipment or procedures.

4.7.4 Test

Test is an element of verification or inspection that generally denotes the determination, by technical means, of the properties or elements of items, including functional operation, and involves the application of established scientific principles and procedures. The product item is subjected to a systematic series of planned stimulations often using special test equipment.

Performance is quantitatively measured either during or after the controlled application of real or simulated functional or environmental stimuli. The analysis of data derived from a test is an integral part of the test and may involve automated data reduction to produce the necessary results.

4.7.5 Special

The special method refers to a combination of some of the other four methods, entailing the use of some modeling or complicated computer simulation mixed with analytical work, or any other verification method that is hard to classify. Many consider simulation as part of the analysis method.

4.7.6 None

No verification action is needed. No compliance evidence need be prepared or submitted. This method will be applied commonly when the paragraph is composed only of a paragraph number and title with no text. This case often is appropriate as a means to organize content. Another example of a requirement that need not be verified is one that simply provides a definition of a term used in the specification. It is important that customer and contractor reach an early agreement on the content of program specifications that will require no verification.

4.8 PRODUCT AND VERIFICATION LEVELS

Figure 28 includes a column titled "Level" to identify the product entity level of hardware or software at which a requirement should be verified. Generally, we will conclude that a requirement should be verified at the item level – that is, at the level corresponding to the specification within which the item requirement and traceability matrix appears. This means that we would run a set of test, examination, demonstration, analysis, and special tasks focused on each item covered by a specification. Other possibilities are that we could do the verification work at a higher or lower assembly level. If the item is a component, we could do the verification work at the subsystem, end item, or system level. Similarly, if the item was a subsystem, we could do the verification work at the component level.

Most often, when the level is other than the item level, it will be a higher level. We will use the words *promote* and *demote* to cover the cases where the verification level is to be accomplished at a higher or lower level, respectively, than the specification in which the requirement appears. This does introduce some complexity into the planning process but does provide the planner with some flexibility that can aid in controlling overall verification cost.

Suppose that we were dealing with the requirements for a liquid propellant rocket engine. We may require the engine supplier to verify the procurement specification liquid oxygen and hydrogen flow rates corresponding to full thrust at sea level in a test prior to shipment and also require that these levels be verified after assembly of the stage, shipment to the launch site, and erection of the complete launch vehicle in a launch pad test. We would mark the requirement in the engine specification for verification at the item level (engine) and at the system level (launch vehicle on the pad). Note that in this case we are dealing with acceptance requirements verification rather than qualification requirements verification. It actually is more common in acceptance that we would isolate verification actions to the item level than in qualification.

The V program model shown in Figure 11 in Chapter 1 offers a very illuminating view of the verification levels applied in product development. At each stage of the decomposition process we are obligated to define appropriate requirements constraining the design solution space, and for each requirement, we have an obligation to define precisely how we will verify that the design solution will satisfy that requirement. This verification action should normally be undertaken at the item level, but it may make good sense for particular item requirements to be verified at a higher level, with requirements for several items being verified in a coordinated series of tests and analyses, the total cost of which would be less than a lot of lower-level work completed on the components that compose the item.

In addition to item qualification and acceptance verification activity, a third level is system test and evaluation. Like qualification, the system test and evaluation work would normally only be done once. Following the completion of the qualification process for configuration or end items, the program should have arrived at a point of some confidence that the system will function properly. It is still possible, of course, that when the verified items are combined into a system they will not perform as specified in the system specification, even though they all passed item qualification. When dealing with very complex systems, such as aircraft, this final form of verification must be accomplished within a very special space and under very controlled conditions to minimize the chance of damage to property and threat to the safety and health of persons. This is especially true if the system in question involves military characteristics that are commonly very destructive, especially when everything works properly.

In the case of aircraft, we call this activity *flight test*. The aircraft passes through a series of ground tests, including a taxi test where all systems can be proven to the extent possible without actually leaving the ground. This will be followed by actual takeoff and landing exercises and flights of increasing duration, increasingly demanding maneuvering trials, and gradually expanding airspeed-altitude envelope excursions with increasingly demanding loading conditions. Similar graduated testing work is accomplished on land vehicles, ships, and space vehicles.

It is possible that an ongoing acceptance kind of verification may also be applied on a program. If the product enjoys a long production run with multiple blocks of product, as in a missile program, each block may have to be tested in a flight test program to ensure a condition of continuing repeatability of the production process. Even where the production run is constant, individual product items could be selected at random to undergo operational testing to accumulate statistical data on production quality, continuing tooling trueness, product reliability, and storage provisions adequacy.

Most often, initial system-level testing is accomplished by the builder, sometimes under the supervision of the customer. In DoD this kind of testing is called *development test and evaluation* (DT&E), as previously discussed in subsection 4.2.4. Customers may conduct their own form of these tests once the product has completed DT&E, called *operational test and evaluation* (OT&E). The latter may be broken into *interim* (IOT&E) and *follow-on* operational test and evaluation (FOT&E) activities. Operational testing is commonly accomplished with actual production items, while development testing is accomplished using special production units that will generally have special instrumentation equipment installed to collect and record performance information that can be evaluated during real-time testing as well as subsequent to flight. ISO/IEC 15288 refers to customer system testing to user requirements as *validation*, but the author groups it together with DT&E as *verification* at the system level.

4.9 VERIFICATION REQUIREMENTS DEFINITION TIMING

The right time to define verification requirements is at the same time as the product item requirements are defined. The reason for this is that it encourages the writing of better item requirements. It is very difficult to write a verification requirement when the item requirement is not quantified, because you cannot clearly define an acceptable pass-fail criteria. For example, how would you verify a requirement stated as "Item weight shall not be excessive"? On the other hand, if the requirement were stated as "Item weight shall be less than or equal to 158 pounds," it can be very easily verified by weighing the item using a scale with a particular accuracy and determining if the design complies.

Ideally, all requirements analysis work should be accomplished by a team of specialists supporting one item specification principal engineer. The principal, or person assigned by the principal, should accomplish the structured decomposition process for the item and assign responsibility for detailed requirements work, such as value definition. For example, the principal would identify a need for a reliability requirement directly from the specification template and depend on the reliability engineer to provide the failure rate allocated to the item using the system reliability math model. Whoever accomplishes the detailed analysis work should be made responsible for determining how the requirement will be verified and for writing one or more verification requirements. Given that the verification method is test, a test engineer may have to be consulted in writing the verification requirement. If any other method is selected, there probably will be no organized functional department that focuses on the method, so the engineer responsible for the specific requirement will likely have to write the verification requirement(s) without the support of a methodological specialist.

4.10 VERIFICATION REQUIREMENTS ANALYSIS

The verification traceability matrix should be fashioned as the requirements are identified. When you are using a modern requirements database, this is done automatically. Otherwise, you will have to create a separate table in a word processor, spreadsheet, or database and suffer the potential problems of divergence between the content of the specification and the verification traceability matrix. The verification requirements must be developed in combination with the item requirements identification. Ideally, the person who writes an item requirement should immediately write the corresponding verification requirement or requirements and complete the other matrix entries as noted earlier. The rationale for this is that it may be very difficult to write a good verification requirement for a badly stated product requirement, possibly encouraging some improvement of the product requirement.

4.10.1 Selecting the Method

The verification requirement must be written in the context of the planned verification method. So, for each requirement in the product requirements sections of a specification, we have an obligation to define how we shall prove that the design solution is compliant with that requirement. How shall we proceed to define that verification requirement in the context of a method? There are two routes that we might consider:

1. Full evaluation. Determine how we can verify the requirement using each of the five methods, and select the one method and its corresponding verification requirement based on a predetermined criterion oriented toward cost and effectiveness.

2. Preselect the method. Based on experience, select the method you think would be most effective and least costly and develop the verification requirement based on that method.

The full evaluation method appears to be a very sound systems engineering approach, involving essentially a trade study for each requirement. Before we eagerly adopt this alternative, however, we should reflect on the cost and the degree of improvement we should expect over the preselect method accomplished by an experienced engineer. Given engineers with experience, the preselect method can be very successful at far less cost than the full evaluation approach. The experience factor can be accelerated by applying written criteria for selection of a method for a product requirement. As suggested in Figure 30, perhaps the first choice should be made between test and something other than test, in that test evidence tends to be very credible.

The first element in that criterion should be to pick the method that provides the best combination of good evidence and low cost. A good rule is to pick the method that is the most effective unless it is also the most costly. In that string, determine the value of the effectiveness as a means of making the final selection. In the case of software, select test unless there is some overwhelming reason not to do so.

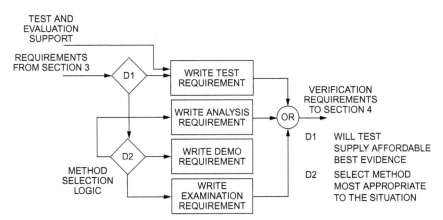

Figure 30 Method selection criteria.

4.10.2 Writing Responsibility and Support

It is a rare person who can do all of the requirements analysis and specification writing work without assistance from specialists in several fields.
This is especially true for verification requirements. A principal engineer
should be assigned for each specification, and that person should ensure
that all of the verification requirements are prepared together with the item
requirements by those responsible. If the test method is selected, someone
from test and evaluation should assist the engineer responsible for that
requirement in writing the corresponding verification requirement. For
other methods, the responsibility for writing the Section 3 requirement
should simply flow through to the responsibility for writing the verification
requirement. In the case of specialty engineering requirements, the specialty engineer can be depended upon to provide self-help in the areas
of analysis, demonstration, and examination. For example, if the requirement was crafted by a reliability engineer, that reliability engineer should
provide a requirement for verifying that requirement, perhaps through
analysis involving prediction of the reliability from a piece parts analysis
or via reliability testing. This same engineer may need help from a test
engineer in the case of test verification.

4.10.3 Writing the Verification Paragraph

Given a particular method has been determined or is being considered for a
particular item requirement, we must conceive what results will satisfy us
that the product design clearly does or does not satisfy the item requirement.
We wish to define the requirements for the verification process, not actually
write the procedure, which should unfold from the requirements. The verification requirements should constrain the verification procedure design as
the item requirements constrain the item design. However, you may find
this very difficult to accomplish in the process of writing the verification
requirement for a product requirement. This is a signal that the requirement
is not well stated and you should spend some time thinking about its restate-
ment. It will most often be helpful to ask someone else to read the paragraph
and offer critical comment. Some examples of verification requirements
coordinated with product requirements are offered in the following
paragraphs.

4.10.3.1 Example of Qualification Examination for Compliance

Note that we have accounted for the possibility that when the scale reads 133
pounds and 13 ounces (134 pounds less half of the worst case plus or minus
error figure), it is actually responding to 134 pounds because of inaccuracy in

the scale. Depending on how critical the weight is, one can select a scale with a finer accuracy, of course.

"3.1.19.3   Weight
Item weight shall be less than or equal to 134 pounds.
4.4.5   Weight
Item weight shall be determined by a scale, the calibration for which is current, with an accuracy of plus or minus 3 ounces. The item shall be placed on the scale located on a level, stable surface and a reading taken. Measured weight shall be less than 133 pounds and 13 ounces."

4.10.3.2 Another Example of Qualification Examination for Compliance
Note that a qualification test could also be conducted to demonstrate durability of the markings under particular conditions.

"3.1.13   Nameplates or Product Markings
The item shall be identified with the item name, design organization CAGE code, and part number using a metal nameplate in a fashion to survive normal wear and tear during a lifetime of 5 years.
4.4.42   Nameplates or Product Markings
Inspect the item visually for installation of metal nameplate and assure that the nameplate includes the correct item name, CAGE code, and part number applied in a fashion that will remain readable under anticipated wear and tear over a five-year period."

4.10.3.3 Example of Qualification Analysis for Compliance
Note that we have allowed for an analysis error margin of 100 hours.

"3.1.5   Reliability
The item shall have a mean time between failure greater than or equal to 5000 hours.
4.4.32   Reliability
An analysis of the final design shall be performed to determine piece part failure rate predictions and the aggregate effects accumulated to provide an item mean time between failure figure. Predicted mean time between failure must be greater than 5100 hours in order to ensure that any error in prediction will not result in failure to meet required value."

4.10.3.4 Example of Qualification Test for Compliance
Note that design faults could pass through our verification process. Faults could exist that result in this mode being entered from an improper mode. We have tested for improper exit but not for improper entry.

"3.1.3.5   North-South Go Mode
The item shall be capable of entering, residing in, and exiting a mode signaling North-South auto passage is appropriate, as defined in Figure 3-5 for source and destination states and Table 3-3 for transition logic.

4.4.15   North-South Go Mode

The item shall be connected to a system simulator and stimulated sequentially with all input conditions that should trigger mode entry and mode exit and the desired mode shall be entered or exited as appropriate. While in this mode, all possible input stimuli other than those that should cause exit shall be executed for periods between 100 and 1000 milliseconds resulting in no mode change. Refer to Figure 3-5 and Table 3-3 for appropriate entry and exit conditions."

4.10.3.5 Example of Qualification Demonstration for Compliance

In this verification requirement, one could criticize the way the demonstrator is to be selected, since it is possible that a tall person could have small hands.

"3.1.17   Human Factors Engineering

3.1.17.1   Internal Item Access

All nonstructural internal items shall be accessible for removal, installation, and adjustment.

4.4.38   Internal Item Access

A man shall be selected who is at or above the upper extreme of the 95th percentile in stature (73 inches) and that person shall demonstrate removal and installation of all parts identified through logistics analysis for removal and replacement and unpowered simulation of any adjustments in the item."

4.11 SPECIAL PROBLEMS WITH INTERFACE VERIFICATION

Interface entities have a unique character relative to product entities in that they have two terminals, each commonly attached to a different terminal product entity. Personnel involved with each terminal should participate in the verification work accomplished in interface verification tasks. This can get more complicated, for example with a wire harness that connects five different "black boxes." The plan for qualification of the design would do well to have some participation by engineers involved in each of the five product entities. A similar case could involve one unit providing fluid to four others under various pressure and flow rates in those four loads. Clearly there are integration and optimization actions that should have taken place during the design in these cases, but also shared responsibilities during verification as well.

CHAPTER 5

Establishing and Maintaining Enterprise Readiness for Program Implementation of Verification

5.1 OUR OBJECTIVE IN THIS CHAPTER

Previous chapters have provided an introduction to verification work that must be accomplished on system development programs. Our objective in this chapter is to identify the resources an enterprise must be prepared to deploy to a new program to place that program in a state of readiness to successfully accomplish verification work. The enterprise through its functional departments should maintain these resources so as to deploy them as needed by a new program. Upon receipt of these resources, the new program must accomplish any adjustments demanded by the contract and acceptable to enterprise management. Where there is a disagreement on the use of these resources, the program manager must resolve the issue as quickly as possible, making sure the customer understands that the enterprise wishes to deliver the same high quality work to all of its customers and that a standard set of resources and processes is an important part of that expectation. Many of the resources for verification work are in the form of templates for deliverable documents and the enterprise can reduce the amount of possible conflict by understanding the preferences held by common customers.

Ideally on every new program the resource base will flow into the program as a function of the personnel assigned to the program from the functional departments reporting in to the program on their first day. But it may be necessary for program management or systems engineering personnel to seek out appropriate functional department personnel for access to needed work products if they are not forthcoming.

It is generally inappropriate for a program to accomplish program work normally assigned to a particular department without that department having been assigned a program budget for doing that work. The department has an obligation to provide personnel who know what they are doing

System Verification
http://dx.doi.org/10.1016/B978-0-12-804221-2.00005-X

and the raw materials needed to support that work. There may be special cases where a functional department cannot supply a person to do a particular task during the time it needs to be done, but the program can assign someone from a different department already on the program to do the job using resources from the department not able to supply the person to do the job, but this should be negotiated between the program and functional department managers.

If a functional department fails to deliver an input work product registered to the department, it should be reported by the program to top management and the missing materials condition should be corrected on an urgent basis. Each department should possess all of the materials that a program needs to function from the perspective of that department and be prepared to turn over a set in very short order (one work day or less). The author introduced the idea of an enterprise integration team (EIT) in a system engineering management book several years ago, motivated by this very concept. The EIT should prepare a list of all of the resources programs require from functional departments and periodically review the state of readiness of the functional departments to provide them to new programs. The EIT should also accomplish integration and optimization work across these materials from time to time and correct any discrepancies discovered that might lead to discontinuities between work accomplished by people assigned to a program from two or more departments.

5.2 DEPLOYABLE RESOURCE RESPONSIBILITIES

In the author's view, it is the responsibility of functional department managers to possess a department manual telling what work the department must be able to accomplish on enterprise programs and how that work is to be accomplished. An enterprise integration team should have the responsibility of integrating and optimizing the department assignments to ensure that all needed program work has been accounted for and that the department manuals comprehensively document department responsibilities. Each functional department manager must have a list of work products people from the department must be capable of developing and delivering on programs. Where these work products are documents, the department should possess a template and a data item description (DID) that tells how to transform a template into a final work product on a program.

5.3 NEW PROGRAM RESOURCE BASE

The functional departments of an enterprise should maintain a complete collection of resources any new program will require such that these resources may be brought aboard the new program as needed. Many of these resources will be compatible with computer storage, so transfer to a program will be a simple matter of downloading.

5.3.1 Personnel

The primary functional department in verification work is system engineering. A system engineer should be assigned as the overall verification activity principal engineer reporting to the program integration team leader, with easy access to the program manager. The systems engineering department must therefore retain sufficient department members who are qualified to do this work. Major product cross-functional teams may also require assignment of verification principal engineers during portions of the program. Ideally, persons from programs should be able to discuss planned assignments from departments to find out which persons would be recommended for leadership roles.

Personnel from many other functional departments will have to act as verification task principal engineers on programs and these departments will have to ensure that their personnel are qualified to do this work. This work will include preparing plans and procedures, accomplishing a planned verification task in their domain, and reporting the results.

All personnel assigned to a program by a functional department must be qualified to do the work described in their functional department's manual. Functional department managers must find affordable ways to cause their department personnel to reach this level of competence by hiring good people, encouraging them to become familiar with the contents of the department manual, and through their attendance in whatever training sessions the budget or employee tolerance for attendance during lunch hour or after work sessions will support.

5.3.2 Document Templates and Data Item Descriptions

An enterprise should seek to standardize its work products to the extent possible. The value in so doing is that employees will gain valuable experience on every program, encouraging more efficiency, lower cost, and tighter time schedules on all future programs. An enterprise that serves a single customer may be able to standardize using the customer's standards, but an enterprise that deals with multiple customers will have to reach its own best choice, encouraging compliance with customer standards while also complying with enterprise conclusions about good business practices.

The work products related to verification tend to be documents of several kinds. For each document work product, the enterprise should maintain a template that is simply a preferred paragraph structure providing essentially an outline of the document. Also the enterprise should maintain for each document work product a document referred to as a *data item description* (DID) that tells how to transform a template into a complete work product of a particular kind for a program.

An enterprise can benefit from retaining several examples of these documents from its history on past programs where the examples were considered having been well done. All of these documents (templates, DID, and samples) being computer generated can, of course, be stored in some form of library from which programs can view and download content at will. The total of library content will necessarily have been contributed by many different functional departments, so once again the EIT should be made responsible for ensuring that all of these departments are current in their responsibilities and that the materials they have provided satisfy any quality requirements the enterprise insists on compliance with.

5.3.2.1 Specification Templates and DID
The specifications that drive the whole verification activity would be more properly listed in a book focused on architecture and requirements engineering, but the work is so tightly linked to verification work that the author has chosen to list them in this book. Figure 31 lists these document templates

System Architecture Report Template and DID
System Specification Template and DID
Item Performance Specification Template and DID
Item Detail Specification Template and DID
Interface Performance Specification Template and DID
Interface Detail Specification Template and DID
Part Specification Template and DID
Material Specification Template and DID
Process Specification Template and DID
Specification Change Notice Template and DID

Figure 31 New program specification and requirements analysis templates and DIDs.

and DID and the reader should recognize that each line item corresponds to two standard documents, a template and a DID. By the time the program begins verification work, these templates will already have been provided to the program and used as the basis for development of program specifications.

Chapter 2 explains the kinds of templates and DID documents related to program specifications of the several kinds. Chapter 3 provides an insight into the kinds of content that would be contained in the document identified as a System Architecture Report. The exact content will be a function of the universal architecture description frameworks (UADF) selected as the basis for determining the system architecture and the requirements derived for inclusion in program specifications.

5.3.2.2 *Verification Document Templates and DID*
Verification is a very document-centric activity. Figure 32 illustrates the kinds of documents needed. The blocks that are colored yellow (light gray in print

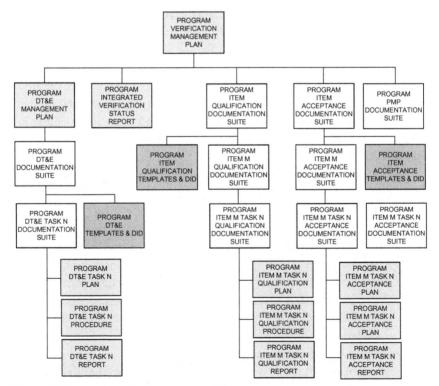

Figure 32 Program verification document tree.

versions) are actual program documents that will have to be prepared on every program. Blocks colored dark gray represent templates and DIDs that will have to be supplied to every program from the responsible functional department, systems engineering more than likely in this case. The white blocks are organizing blocks. The enterprise should provide programs with templates and DID for the program verification management plan, program development test and evaluation (DT&E) management plan, and program integrated verification status report, but it was not convenient to show those on the figure.

Figure 33 provides a detailed list of templates and DID for the other three classes recognizing: (1) plan, (2) procedure, and (3) report. An enterprise can choose to combine the first two into a single document with no loss in verification effectiveness. The reader should note that at the system level we can deal directly with verification tasks, but at the item level we have to deal first with the item involved (M) and then the verification tasks (N) for that item.

Figure 33 separates these three kinds of documents with a blank line to emphasize the distinction. The item qualification and item acceptance collections should be identified as item and interface qualification and acceptance, but in the interest of reducing the length of the title names the author has left off the interface relationship in Figure 33. The reader will note that these documents fall into the same pattern for three classes of verification: (1) item qualification, (2) item acceptance, and (3) DT&E and five methods of verification: (1) analysis, (2) test, (3) examination, (4) demonstration, and (5) special. There is, of course, a sixth class referred to as none, where no verification is required but there is also no documentation we need associate with this class.

The PMP tree is not shown expanded in Figures 32 and 33 but the reader can imagine that it includes a trio of documents: (1) parts, (2) materials, and (3) processes. Under each of these headings will be plans, procedures, and reports. In many cases the parts, materials, and processes (PMP) documentation will involve single sheets but in others considerable data. The author is familiar with one program where the file cabinets containing all of the supplier data forced all of the program data management people out of the space where it was stored in file cabinets before the program switched to computer data. Computer media is a better alternative than paper in folders, as it is in all verification documentation cases actually. PMP is also omitted from the listing in Figure 32.

The plan for a task tells how the task principal shall proceed to complete the task and what resources will be needed. The procedure tells in some

Item Qualification Verification Analysis Task Plan Template and DID
Item Qualification Verification Test Task Plan Template and DID
Item Qualification Verification Examination Task Plan Template and DID
Item Qualification Verification Demonstration Task Plan Template and DID
Item Acceptance Verification Analysis Task Plan Template and DID
Item Acceptance Verification Test Task Plan Template and DID
Item Acceptance Verification Examination Task Plan Template and DID
Item Acceptance Verification Demonstration Task Plan Template and DID
System DT&E Analysis Task Plan Template and DID
System DT&E Test Task Plan Template and DID
System DT&E Examination Task Plan Template and DID
System DT&E Demonstration Task Plan Template and DID

Item Qualification Verification Analysis Task Procedure Template and DID
Item Qualification Verification Test Task Procedure Template and DID
Item Qualification Verification Examination Task Procedure Template and DID
Item Qualification Verification Demonstration Task Procedure Template and DID
Item Acceptance Verification Analysis Task Procedure Template and DID
Item Acceptance Verification Test Task Procedure Template and DID
Item Acceptance Verification Examination Task Procedure Template and DID
Item Acceptance Verification Demonstration Task Procedure Template and DID
System DT&E Analysis Task Procedure Template and DID
System DT&E Test Task Procedure Template and DID
System DT&E Examination Task Procedure Template and DID
System DT&E Demonstration Task Procedure Template and DID

Item Qualification Verification Analysis Task Report Template and DID
Item Qualification Verification Test Task Report Template and DID
Item Qualification Verification Examination Task Report Template and DID
Item Qualification Verification Demonstration Task Report Template and DID
Item Acceptance Verification Analysis Task Report Template and DID
Item Acceptance Verification Test Task Report Template and DID
Item Acceptance Verification Examination Task Report Template and DID
Item Acceptance Verification Demonstration Task Report Template and DID
System DT&E Analysis Task Report Template and DID
System DT&E Test Task Report Template and DID
System DT&E Examination Task Report Template and DID
System DT&E Demonstration Task Report Template and DID

Figure 33 New program verification documentation templates and DIDs.

detail how each step shall be accomplished. The report in each case gives the results of having accomplished the task. The latter states the evidence of the degree of compliance actually achieved in the design relative to the requirements based on the work accomplished in the verification task.

Figures 32 and 33 provide a very expansive listing of documents that can be trimmed to a minimized list that many enterprises will find perfectly adequate for program purposes. First, it is possible to merge the content of the plan and procedure into a single plan document resulting in only a set of plan and report documents. The procedural data is simply contained in a set of steps in the plan. Second, one can delete the interest in data item descriptions or delay interest in them until the set of templates is mature. Third, the templates for plans and reports can be developed to cover all five methods rather than prepare one for each method. The result of these actions reduces the number of templates one would have to prepare to those nine shown in Figure 34. Granted, these templates would have to include a lot of guidance for covering the many applications of each template. Appendix A includes a set of the templates shown in Figure 34.

The program verification management plan is intended to be the overall plan for all verification work accomplished by the program and all other verification documentation should be subordinate to it. The System DT&E is a very important part of the overall activity and should be documented separately for program and customer review and approval and identified in the program verification management plan. The verification status report contains the verification program man-hour and schedule assignments,

A1	System DT&E and Item or Interface Qualification Verification Task Plan Template
A2	System DT&E and Item or Interface Qualification Verification Task Report Template
A3	Program Functional Configuration Audit Report Template
A4	Item or Interface Acceptance Verification Task Plan Template
A5	Item or Interface Acceptance Verification Task Report Template
A6	Program Physical Configuration Audit Report Template
A7	System DT&E Master Plan Template
A8	System DT&E Report Template
A9	Program Parts, Materials, and Processes Master Plan

Figure 34 New program simplified verification documentation templates.

identifies all tasks and their principal engineers, contains a set of matrices needed in management of the overall process: (1) verification compliance matrix, (2) verification task matrix, and (3) verification item matrix. The report should be reissued periodically providing status in the three matrices. Ideally, these matrices would be available in real time from supporting computer resources. This document provides management with the information they need to manage the verification effort.

5.3.2.3 An Omitted Document

Commonly, a program will require performance of a DT&E by the contractor under the contract that is intended to determine the extent to which the system complies with the content of the system specification. Often, the customer will insist on accomplishing an operational test and evaluation (OT&E) at some point after the DT&E is complete and the customer has taken delivery of the low rate production end items and spares needed to support the testing, and sufficient customer personnel have been trained to accomplish the work in accordance with the content of validated and verified technical data delivered. The OT&E is commonly implemented in accordance with an OT&E Plan crafted by the user and accomplished by the user to determine the extent to which the system complies with mission needs the user intended to employ the system to achieve. This document and the lower tier plans, procedures, and reports are not included in Figures 32 and 33 because they are not a contractor responsibility.

5.3.3 Requirements and Verification Database

Programs should be served by an effective computer database, into which all of the content of all of the program specifications can be entered, including, of course, all of the product and verification requirements. This database may include modeling capabilities from which the product requirements are derived or the related sketches may be prepared by other means and referenced in the database for traceability purposes. The modeling sketches may even be hand drawn on paper with pencil, but in any case these modeling work product masters should be placed under configuration control just as the specifications and verification documentation should be.

The database should support verification status and management needs, providing views and/or reports of the several tables or matrices discussed in this book. The system should be capable of publishing all program specifications and all verification plans, procedures, and reports. The masters for every document published may be retained in the requirements and verification database system or as separate documents under configuration management control but all versions of every document must be available throughout the program life span.

Figure 35 suggests a tabular structure of a federated database deployed to programs by some combination of functional departments, such as system engineering, test engineering, and specialty engineering. Product requirements are captured in a requirements table, linked to particular methods of verification and particular verification requirements captured in a verification requirements table with traceability clearly identified. Strings are coupled to tasks for which plans and procedures are created and the resultant evidence linked to report content.

Ideally, a computer database to support requirements and verification work would employ a primitive requirements capture concept where the table into which all requirements were placed would capture the attribute (ATT), relation (REL), value (VAL), units (UNIT) such that the computer could actually write the paragraph for inclusion in the final specification. A specification sentence could be generated in the language of choice by the following line of code in a database application: PARA+" "+"Item "+TRIM(ATT)+" shall be "+REL+" "+VAL+" "+TRIM(UNIT)+".". The code would have to be a little more complex than this because of different cases for kinds of relationships, system versus item, and a few other complexities, but the author built such a database in the 1990s that would do this. The advantages include treatment of the values in a numerical field rather than as part of a text string, and these values could be dealt with in several useful ways using fairly simple arithmetic within the requirements database, as well as cooperatively with other applications.

5.4 ASSURANCE OF PROGRAM COMPLIANCE

An enterprise can prepare the most effective standard practices and resource base as a basis for beginning a new program and can prepare the most clearly written direction for program manager compliance with those standards in implementing the program, but many program managers and some of those reporting to the manager will believe they know better. An enterprise must apply a verification process to itself to ensure that programs comply with enterprise direction in this regard. The best enterprise function for this

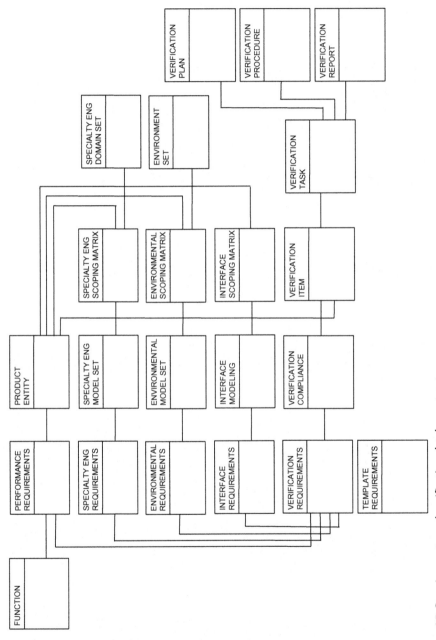

Figure 35 Requirements and verification database structure.

verification work is a quality assurance function. To implement the practice, quality assurance needs only to inspect work being accomplished on the program as soon as possible to determine the extent to which the program is following common process direction. Enterprise management must treat a failure to comply harshly because common process carries with it such tremendously important benefits. More will be said on this in Chapter 11.

CHAPTER 6

Verification Process Design for a Program

6.1 OVERVIEW OF THIS CHAPTER

This chapter begins in Paragraph 6.2 with a discussion of the overall verification process, emphasizing the commonalities between the four classes of verification. The Systems Verification function, identified as F124 in the remaining chapters of the book, depicted as F44 on the system life-cycle flow diagram shown in Figure 7 and F144 on Figure 21, is expanded in preparation for further description of the process in detail in this chapter and further still in the four class-focused Chapters 7, 8, 9, and 10. A common error in verification work on a program is a failure to execute a two-stage design of the verification process based on a set of requirements appearing in program specifications. Paragraph 6.3 summarizes how to do this work well. As in all systems engineering work, while we may identify work in unique isolated tasks, it is necessary to integrate and optimize across these many tasks to reach a really successful plane of performance. Paragraph 6.4 addresses this reality and includes a simple technique that will guarantee success.

6.2 SYSTEMS VERIFICATION PROCESS OVERVIEW

Even though verification work comes after requirements and synthesis work on a program, as shown in Figure 36a, planning for the verification activity must begin early in the program in order to ensure that the program possesses the resources and information needed when verification work actually begins. Ideally, an enterprise would apply a common process to the whole system development activity, including verification, and would therefore have to pass through much of the work discussed in this chapter and the prior chapter for the first time only once, with minor improvements made over time based on program experiences. In this case, an enterprise systems engineering manual would provide guidance for programs on implementing a standard verification process and this process would be embedded in the program planning. Figure 36b shows a simplified expansion of the verification work expressed as function F44 on Figure 7.

System Verification
http://dx.doi.org/10.1016/B978-0-12-804221-2.00006-1
155

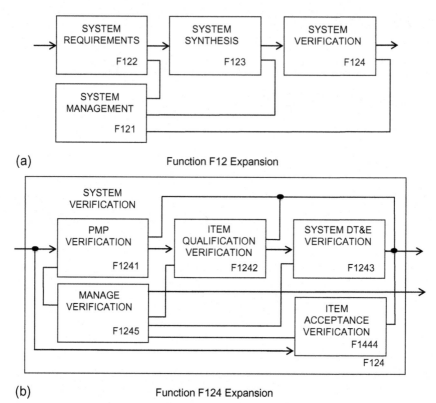

(a)   Function F12 Expansion

(b)   Function F124 Expansion

Figure 36 Simplified Functional View of the System Development Process.

6.2.1 System Level Planning

As soon as a program is established, work should begin on three verification documents in the Manage Verification function (F1245): (1) program verification management plan, (2) program development test and evaluation (DT&E) management plan, and (3) program integrated verification status report. The work that can be accomplished on these documents will not be substantial early in the program, but the structure of the documents can be created and program policy stated. The program verification management plan is the top-level program plan for verification and much of the content can be fairly generic from one program to another. The program DT&E management plan, which some people might refer to as the program system test and evaluation master plan (TEMP), provides details that should be summarized in the program verification management plan. The program level plan can cover the item qualification, item acceptance, and parts, materials and processes (PMP) activity planning in some detail, but a customer will often insist on a separate DT&E plan.

The program integrated verification status report will depend on an effective program database covering requirements and verification data to permit timely reports of program status as the program moves into verification work. A program will have plenty of time to lay out the structure of the report because the earliest information available for this document will start becoming available in terms of PMP selections during design work. The key content of this document identifies all verification tasks, the principal engineer assigned in each case, task budget, task schedule, and status. It is through the information contained in this document that managers have some chance at success in managing the verification work on a program.

6.2.2 Verification Planning During System Requirements Work

Two important verification planning activities occur during the requirements work leading to publishing specifications for product entities. Both take place during efforts to prepare the verification traceability table appearing in each specification. This table, a fragment of which is shown in Figure 28, lists every product requirement followed by the corresponding verification requirement and two choices. First we must decide what method will be employed in verifying the product requirement and, secondly, at what level the requirement will be verified. We have to choose between test, analysis, examination, demonstration, special, and none for the method. Most often the level will be at the level corresponding to the specification in which the requirement appears but there may be cases where the program could be better served by promoting the verification work to the parent item, or demoting it to a child item level, as discussed in Chapter 4.

There are verification planning activities appropriately done from the top down and those best done from the bottom up. This chapter provides a view of the whole as if we were looking in from the outside. Figure 36a provided a program and system life-cycle functional flow diagram and the intent in that kind of diagram is to expose how a process is accomplished in terms of a number of functions or tasks interconnected in a particular sequence. Each block in the diagram can be explained in text, a set of needed input resources listed, output work products created listed, and, if necessary, the flow diagram decomposed into a lower order of granularity with an expansion of the descriptive material. The companion Elsevier book to this book, *System Requirements Analysis*, explores function F122, Systems

Requirements, and this book picks up the story exploring Function F124 of that diagram, titled Systems Verification. In this book we will apply the functional model for a verification common process description, but a devotee of MSA-PSARE or UML-SysML can refer to the functional bubbles of MSA-PSARE or the activity diagrams of UML-SysML as functions with little conflict in meaning.

This chapter offers a decomposition of function F124 from the top down, to show the four major classes in which it is accomplished, and identifies the verification management function. But before exposing that formal decomposition of the verification flow diagram, let us view the basic process that will take place across all four classes and for each specification of the several kinds. Figure 37 exposes a summary view of the top-down versus the bottom-up verification work. Clearly the verification work on a program will have little chance of success if the architecture and requirements engineering work is poorly accomplished on the program. All program specifications including system and entity flow into the top-down verification work. The entity specifications should be accepted as including the item, interface, parts, materials, and processes specifications.

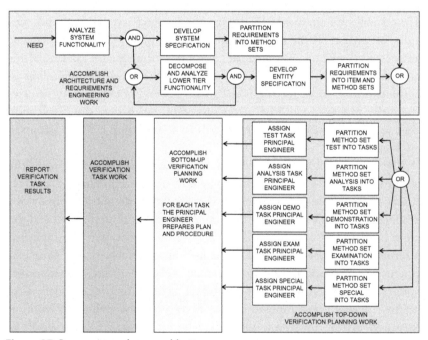

Figure 37 Program top-down and bottom-up processes.

Every program specification can be treated the same way in the top-down process.

The top-down verification work begins with each specification entering the verification work containing product and verification requirements linked in a verification traceability matrix with the method that will be applied to verify design compliance and the level at which it should take place. Each specification should include this traceability matrix. System engineers assemble the content of the union of all of these matrices appearing in all program specifications into a program-wide verification compliance matrix. The content of this matrix is used to partition all of the records in the matrix into a finite number of tasks, each applying one of five methods: (1) test, (2) analysis, (3) examination, (4) demonstration, and (5) special. The responsibility for each task is then assigned to a principal engineer, who must accept the responsibility to develop a plan and procedure, bring together the material and personnel resources needed to accomplish the work, perform or manage the performance of the work defined in the plan and procedure, and prepare a report telling to what extent the product complied with the requirements assigned to the task. The assignment of principal engineers for tasks completes the top-down planning work for a program and paves the way for the bottom-up work to begin in the form of the task principal engineers building their plans and procedures in accordance with a schedule and budget assignment.

The top-down process description for a particular program starts with the availability of verification traceability matrices in the program specifications (system, item performance, and item detail, interface performance, interface detail, parts, materials, and processes). The design of the verification work corresponding to these specifications can begin at this point. It is actually not necessary for a program to wait until all of the specifications are complete with these verification traceability matrices to begin the bottom-up planning process. As the system specification is reviewed, approved, and published, planning can begin for system test and evaluation, but a program is well advised to go slow on the detailed task planning, focusing primarily on the overall DT&E plan until the design of major end items has reached a level of maturity at which risk is believed to be clearly under control.

As each item performance specification is completed, the program can complete the identification of tasks and selection of task principal engineers for item qualification, but the bottom-up work of plan and procedure development should be delayed until the top-down work has been completed for the subordinate entities, because of the possibility that the team will conclude that some promotion and demotion may be appropriate. In these cases, it is decided that verification of particular requirements for item A134, for example, can best be determined through promotion to tasks accomplished on item A13 or demoted to tasks accomplished on items subordinate to A134 such as A1341. Parts, materials, and processes verification work must await selection of these entities in the design process and identification of preferred sources. Bottom-up planning for item acceptance verification is best set aside until the design of the product matures to some extent and the detail specification is complete.

6.2.3 First Expansion of Life-cycle Function F124

Figure 38 expands on the F124 function of Figure 36a and additional flow diagrams will expand the blocks on Figure 38. In this chapter we will discuss preparatory steps to provide the program with an important infrastructure to support an organized implementation of verification at the part, material, process, item, interface, and system levels. Table 1 lists the five major functions subordinate to function F124, tells the names of these functions, identifies in what chapters of this book they are discussed, and identifies the paragraph in this chapter where the function/class is briefly discussed. Function F1245 includes all general verification management functions including preparatory work, program level planning, verification integration and optimization, and database services. General supervision of the overall process is, of course, also included.

Development evaluation testing (DET) does not show up in this diagram, because the author groups it under validation, along with all of the analyses that evaluate the relationship between item requirements and design alternatives, for the purpose of reducing program risk and gaining confidence that the current design approach will be successful. This work occurs partly in the system requirements function (F122) and partly in what the author calls systems synthesis (F123).

6.2.3.1 Parts, Materials, and Processes Verification

PMP verification on a program (function F1241 on Figure 38) tends not to be as formally accomplished as the other three verification classes – although perhaps it should be. The need for specific parts, materials, and processes is commonly driven by the design engineers who create the design drawings making entries on these drawings calling for the need for particular PMP. An

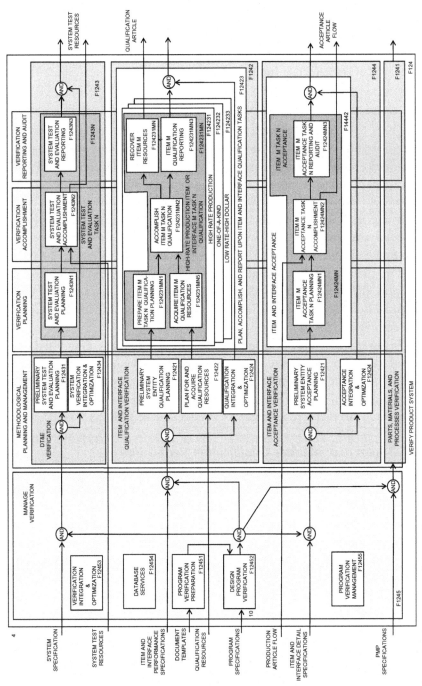

Figure 38 Systems Verification Process.

Table 1 Verification Function/Class Coordination

Function	Function name	Chapter	Paragraph
F1241	Parts, Materials, and Processes	7	6.2.3.1
F1242	Item and Interface Qualification	8	6.2.3.2
F1243	System Test and Evaluation	9	6.2.3.3
F1244	Item and Interface Acceptance	10	6.2.3.4
F1245	Manage Verification	5, 6	

enterprise could organize its functional department structure to include parts engineering in the electrical engineering department and materials and processes department in the mechanical engineering department, or it could group them all together reporting to the engineering manager.

In any case, an enterprise should require all programs to respect an approved parts, materials, and processes list. This list calls for all design engineers to use only items from the list where the enterprise has determined that there are specific suppliers capable of producing and shipping items to the enterprise that will comply with product requirements. So, appearance on the list is essentially a result of the enterprise and its programs having verified item compliance for use on programs.

On a program where a design engineer cannot find anything listed that will comply with program requirements, or prefers the use of an alternative to items listed, the engineer has the responsibility to apply for the item being added to the list. The item is evaluated and a ruling made to add it or to reject the application. Refer to Chapter 7 for coverage of PMP.

6.2.3.2 Item and Interface Qualification Verification

The item qualification process is driven by the item development requirements contained in performance specifications. Item qualification test and demonstration work must be accomplished on test articles representative of the final design, fabricated in a fashion as close as possible to the final manufacturing process as can be arranged, but this work is often actually done in an engineering laboratory by engineers and technicians before a program manufacturing capability is available. All of the item qualification work should be complete as a prerequisite to system testing, where we seek to provide evidence that the whole system satisfies the system specification content. The rationale for this is safety as well as providing for integrity of the chain of verification evidence. It is unsafe to operate the major end items of a system until the parts of them have been qualified for the application. On a DoD program, the results of all of the item qualification tasks will be audited at one or more *functional configuration audits* (FCAs), which may be accomplished in a single audit or a series of item-oriented audits capped by a system FCA. Item qualification is covered in detail in Chapter 8 of the book.

The qualification verification work covered in Chapter 8 is extended to include interfaces between product entities within the system. The requirements driving the design of these interfaces will be covered in terminal specification pairs or interface performance specifications. If an enterprise chooses the latter document pattern, it is recommended that these interface performance specifications be prepared to cover all of the interfaces between a particular pair of product entities. Also, on any program employing interface performance specifications within which to capture interface requirements, they must not double-book these requirements in product entity specification pairs. It will not be possible to preserve integrity between the duplications as changes are made. An effective practice in this case is to include the interface requirements in the interface performance specification, refer to the interface performance specification in the terminal item performance specifications, and list the interface performance specification in Section 2 of the terminal item performance specifications.

Following completion of a successful item qualification action, the program may be required to stage a functional configuration audit where the customer and contractor program personnel meet to formally determine the extent to which the contractor verified that the design complies with the requirements in the item performance specification. As already mentioned, the FCA will be discussed in some detail in Chapter 8.

6.2.3.3 System Test and Evaluation Verification

System test and evaluation takes place on a program when all of the items composing the system end items have passed qualification verification and the end items have passed an evaluation of some key parameter values. We commonly think of qualification applying only to black boxes, but a program would not attempt a first flight of a new aircraft design without the airframe having passed some form of airframe structural integrity test. After all, the airframe is one of the aircraft "black boxes." There are analytical tasks that can be completed prior to flight of an aircraft that are also accomplished at the complete aircraft level. We can, for example, complete the reliability analysis of the complete aircraft through the probabilistic combining of the failure rate figures of all of its parts.

System test and evaluation is often accomplished in two series. First the contractor will carry out a DT&E proving that the system complies with the content of the system specification. Then the customer will take delivery of sufficient assets for the user to conduct an operational test and evaluation (OT&E) to determine if the system can satisfy user needs. One hopes, of course, that the contract and the system specification that was part of the contract were in complete agreement with the user needs, but this is not always the case. We will return to system test and evaluation in Chapter 9 of the book. Often the contractor will have little access to OT&E results in real time and may be unaware that the user is conducting OT&E based on user requirements, some of which were not accepted by the acquisition agent and therefore not included in the system specification included in the contract.

6.2.3.4 Item and Interface Acceptance Verification

The item acceptance process is driven by the item product requirements appearing in item detail specifications and should drive the development of item acceptance verification documentation on manufactured product intended for delivery to the customer. The author prefers that items being manufactured in engineering laboratories in preparation for item qualification pass through an acceptance prior to their use for that purpose. The final acceptance test plans may not be available at this point, but it is an opportunity for engineering to observe the correctness of these preliminary plans and procedures and adjust them toward their final form. It also helps support the story for the customer that the contractor has done everything reasonable to frame the production of the qualification items in a replication of the production item manufacturing experience, supporting the idea that the customer should accept that the evidence gained from qualification item testing should be accepted as representing the evidence that would be obtained from a production item. The first article that flows through the acceptance process will produce evidence of the degree of compliance of the design and manufacturing process with the product requirements. The results of the acceptance process on the first article should be audited at a *physical configuration audit* (PCA) on a DoD program. Subsequently, the acceptance process is applied during or after production of each article or in some sampling pattern. If the article passes, it ships and, if not, it does not ship until observed problems are corrected and it can pass the inspection. Chapter 10 of this book takes up item acceptance verification. The same concern for item detail specification and interface detail specification content, with the same practice to avoid that concern, is encouraged.

6.3 MANAGE VERIFICATION

In addition to these four main verification class functions, Figure 38 includes a manage verification function. During the time requirements analysis activity is in the process of producing specifications for the system and its many items, those responsible for verification must begin to prepare the program for that work in function F12451, covered in Chapter 5. As the reader can see from Figure 38, we have partitioned all of the verification work into many functions, some of which will be occurring simultaneously under the responsibility of numerous people. In any such situation it is necessary to integrate and optimize across all of this work at the system verification level (F12453) and within each verification class. The whole process must, of course, be managed throughout the program run, function F12455. While this work can be accomplished without employing a computer database, no successful argument can be made to support this alternative, so function F12454 is included. Each of the four verification classes also include from one to three management-related functions.

Figure 38 also cross-partitions all verification work into four activities: (1) verification planning, (2) verification accomplishment, (3) verification reporting and audit, and (4) verification methodological planning and management.

6.3.1 Identification of Entities Requiring Verification

In the case of DT&E verification, there is only a single entity involved, the whole system, so all of the DT&E tasks relate to the system specification. If there are fewer than 60 DT&E tasks, they can be identified as F1243N where the character N is assigned uniquely for each task. If there are more than 60 tasks, they can be partitioned into subsets as a function of the facility where they will be conducted or in accordance with some other categorization scheme, such as the method applied. In general, in collecting requirements strings into tasks we are looking for collections of requirements that will require the same supporting resource sets and appeal to the same personnel knowledge domain as well as the same method (test, analysis, demo, examination, or special). In some cases it may make sense to identify a task requiring that people from multiple domains must participate, and in these cases the planners have to recognize that those tasks will require stronger integration and optimization than single domain tasks will require.

Commonly, any PMP task will require a simple task structure and only a single task for any one part, material, or process, as simple as a study of a supplier sheet and a conversation with a person from the supplier company. For a large system, the number of these verification tasks will be very large. Given that the program has an effective preferred PMP list activity, the verification in each case will be to verify that the system application applies no stresses greater than the supplier data covers.

In general, every item identified by management as a configuration item should pass through item qualification and acceptance, but there are possible exceptions. In some cases it is possible that an item could be proven to comply with the content of its specification by accomplishing verification of the parent or all of its child items. Usually, the only motivation for doing so is to save on the cost of verifying the item directly.

Item and interface qualification and acceptance verification are the most complex verification activities in that there will be many of them at different product entity levels of organization and each item or interface will commonly involve multiple tasks for each method. Figure 38 recognizes item M and task N applying to that item. So the complete set of tasks for a particular entity will require identification of some number of tasks for each of the five methods. Generally, the qualification tasks will be accomplished by an engineer assigned as the principal engineer and acceptance tasks accomplished by a person from manufacturing assigned as the principal engineer, but this is subject to how the enterprise chooses to organize and assign responsibility to departments.

For a given entity performance specification or entity detail specification, the verification traceability table will identify the method to be applied for each requirements string as described in the author's previous book, *System Requirements Analysis*. In function F12452 of Figure 38 we would pass from the performance or detail specification for an entity into function F124521 shown in Figure 39, where a decision is made about what entities will be subjected to verification, though commonly these decisions will already have been made while determining what to write specifications on. The verification traceability table entries are transferred into a compliance matrix in the supporting database and will be worked on as a function of the particular verification class of interest: DT&E, PMP, qualification, or acceptance.

Figure 40 expands function F124523 for item qualification. Item acceptance planning is of a similar structure except that it is accomplished on each article of the product flowing through the production process and that the product must retain full life expectancy upon completion of acceptance. Clearly, any program will have to produce many verification documents because each task identified will require a plan, a procedure, and a report.

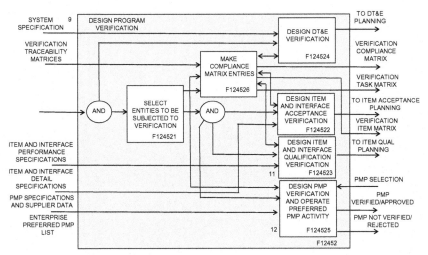

Figure 39 Design program verification.

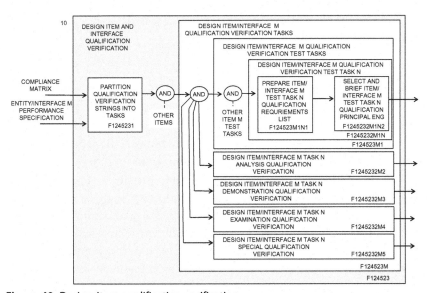

Figure 40 Design item qualification verification.

Figure 41 only expands on the qualification verification for item M and the system may include 103 items requiring qualification. Figure 40 accounts for these other items by referring to "Other Items" that would reflect essentially the same pattern as item M. The diagram accounts for applying some number of tasks for each of the five methods.

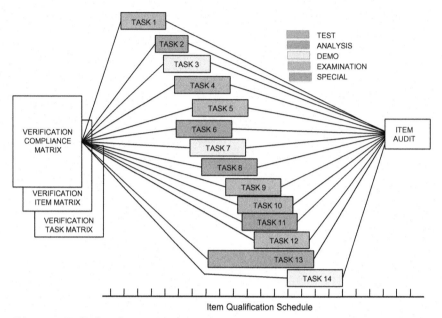

Figure 41 Task identification.

For any given product entity for which a specification has been prepared – whether it be the whole system for which DT&E must be accomplished, an item or interface in a system for which qualification and acceptance must be accomplished, or a part, material, or process for which PMP verification must be accomplished – some number of verification tasks will have to be performed. As mentioned earlier, PMP will often require a single task. For DT&E there is only one entity to verify but the whole system verification process will entail many tasks. Each item and interface qualification and acceptance verification activity will be composed of many tasks involving some combination of the five verification methods, as shown in Figure 41.

The program compliance matrix referred to in Figure 40 includes every verification string for the system and all of its elements. It includes the union of all of the verification traceability matrices appearing in all of the program specifications, with added information to identify the verification task to which it is assigned. The database will have to include tables for items containing status information and tasks containing identification of the principal engineer and task status.

The team responsible for design of the item should complete the task matrix entries including identification of the task principal engineers.

In Figure 40 we see two very important tasks identified to prepare a list of requirements in the compliance matrix, which are assigned to the task and selection and briefing of the principal engineer. The latter must include presentation of the list of requirements to the principal engineer and the instruction that the principal engineer must include this list in the report, coordinated with the exact location in the report where the evidence of compliance is located. One aspect that causes so many programs to experience great difficulty in verification lies in the difficulty those who read/review these verification reports have in finding the evidence of compliance and reaching a conclusion about its veracity. During program review of all verification task reports, the reviewer must observe the table, verify that the location of the evidence in each case is accurately stated and that the evidence can easily be understood, and it is convincing.

6.3.2 Verification Task Identification

In the previous paragraph we danced around a method for identifying verification tasks uniquely, without closing on one or some list of possible methods. A program will need a method for doing so because all of the evidence of compliance will be derived through accomplishment of a task, and we will want to be able to find the report in which that evidence is contained. The author's preferred method involves each task being identified using a string of base-60 symbols collected into three hyphen-separated sets, as illustrated in Figure 42a. The set theoretic figure shows all of the possibilities for a verification task manufacturer's identification number (MID) for one system entity, which is product entity A12. Two other entities are included in the sketch, suggesting the reality that for a particular system this stack would include every product and interface entity including the system, all of the items with specifications, and all of the PMP. This is not to say that every subset shown on the figure would be included for all or any one entity in the system, but the diagram shows every possibility for the formation of a verification task MID.

The reader will note that the author gives precedence to the product or interface identification in the formation of a verification task MID because he believes this sequence will coordinate most beneficially with the intended management approach focused on product entity teams. The next priority is given to the verification class, of which there are the four shown. The third priority is the method chosen to be employed in the task, of which there are five possibilities. Finally, we must identify a unique number within the

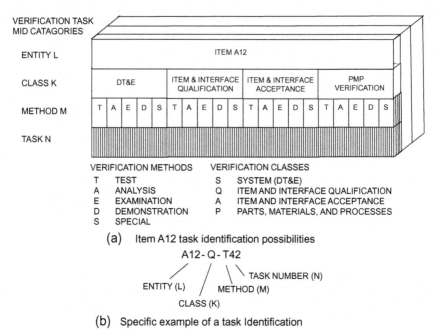

VERIFICATION METHODS
- T   TEST
- A   ANALYSIS
- E   EXAMINATION
- D   DEMONSTRATION
- S   SPECIAL

VERIFICATION CLASSES
- S   SYSTEM (DT&E)
- Q   ITEM AND INTERFACE QUALIFICATION
- A   ITEM AND INTERFACE ACCEPTANCE
- P   PARTS, MATERIALS, AND PROCESSES

(a)   Item A12 task identification possibilities

A12-Q-T42

ENTITY (L)   METHOD (M)   TASK NUMBER (N)

CLASS (K)

(b)   Specific example of a task Identification

Figure 42 Verification task identification possibilities.

subset identified by the entity, class, and method designations. Figure 42b shows an example of a complete verification task MID constructed using the technique encouraged in Figure 42a. This task is clearly associated with some number of requirements contained in the performance specification (the Q class is related to item qualification that is driven by the item performance specification) for product entity A12. The verification method selected is test and this task appears to be the 42nd test task identified in this subset of tasks. Apparently we have chosen to use base-ten numbers in subset N, but we could have chosen to use the base-60 system for the complete task MID, which would have been A12-Q-Th in that "h" is the 42nd base-60 character, as the reader can verify given the complete base-60 character set consisting of the ten Arabic numerals, the uppercase English alphabet characters less the letter *I* and the lowercase English alphabet characters less the letter *l*. Using the base-60 set, we would have to resort to two character N sets if there were more than 60 verification tasks of one class and method for one product entity, but it is unlikely that any program would require so many tasks in one subset string.

6.3.3 Verification Schedule

Figure 43 shows a high-level view of a program verification schedule. The amount of detailed verification work that must be planned and managed is not obvious from a high-level schedule. As an example, the program may be developing a system that requires 35 items that must be subjected to item qualification and acceptance. In this case, the requirements work would involve development of a system specification, 35 performance specifications, and 35 detail specifications during the requirements task. The item qualification work would be driven by the content of the 35 performance specifications, the item acceptance work by the content of the 35 detail specifications, and the DT&E work by the content of the system specification. As the number of items rises to 100 or more in multiple layers, the difficult nature of such a program can be appreciated.

Synthesis design work can proceed for an item once the item performance specification has been approved. The design process will identify parts, materials, and processes that can be worked with those responsible for the standard PMP list, followed by procurement action to check available suppliers. At the item preliminary design review (PDR) the design engineer or team should brief any need for early procurement of parts to satisfy schedule needs for manufacture of qualification articles. Clearly, it will not be possible to manufacture production articles unless the materials identified on engineering drawings have been procured and received so there is some procurement and warehousing work between the end of design for an entity and the end of its manufacture regardless the degree of success in just-in-time delivery attained by the enterprise and its suppliers. The manufacturing process can extend long after the development work that commonly ends at critical design review (CDR).

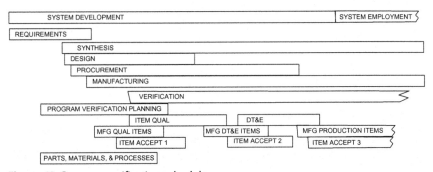

Figure 43 Program verification schedule.

The manufacture of qualification articles can begin after PDR and avail-ability of the engineering lab, materials, personnel, and supporting equipment, as well as the approved plans and procedures for all of the verification tasks that will have to be performed. The item qualification work will consist of some number of verification tasks for each item qualified. Some of these tasks will involve analysis and these tasks will commonly not require access to a product entity to complete. Other tasks involving test, demonstration, or examination will likely require a qualification article. Let us assume for a particular item that all item qualification work can be completed with only a single item. The author maintains that all 35 articles in this example should be manufactured in an engineering shop using preliminary manufacturing instructions and be subjected to acceptance in accordance with the pre-liminary item acceptance plan and procedure before being fed into the qualification work.

For a given item among the 35 required on this program, we may have concluded that it was necessary to accomplish 14 qualification tasks, arranged in a schedule as shown in Figure 44. Each of the tasks illustrated would have to have a plan and procedure developed and some of these tasks would have to have access to a qualification article manufactured to support

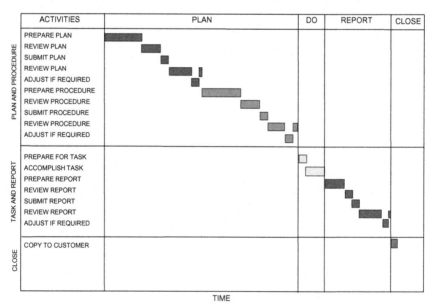

TIME

Figure 44 Typical item verification task schedule.

qualification. Certainly the test tasks would require access and the reader will note that these are shown not sharing schedule time. The demo and examination tasks may require some access to the qualification article as well, and the schedule and related plans and procedures would have to recognize this interaction.

Each task shown in Figure 44 will require a series of three documents prepared for it, including a plan, procedure, and report (unless we accepted the simplified approach combining the plan and procedure in one document). The plan and procedure will have to be completed, reviewed, approved, and placed in protected storage under configuration management control prior to beginning the verification work. The report in each case will have to be prepared subsequent to completion of the task, subjected to review and approval followed by its inclusion in documents under configuration management control. The customer may require that all of these documents be delivered as *contract data requirements list* (CDRL) items and may even reserve approval authority over them in the contract. As a minimum, they will require access to the reports in time to study them prior to a scheduled item functional configuration audit, where they may demand more information supporting the conclusion claimed by the contractor that the design complies with the requirements.

Figure 44 shows a schedule for accomplishing one of the tasks shown in Figure 43. The method applied could be any one of the five noted earlier. The schedule is partitioned into planning wherein the plan and procedure are prepared, doing the work in accordance with the plan and procedure, reporting results, and closing out the task. In the context of Figure 44, this close-out could be as simple as sending the customer a copy of the report and linking it to the item FCA with the documents from the other item tasks.

6.3.4 Verification Budget

When a program is proposed to a customer, it will not be known how much the verification work will cost to perform. An estimate is needed based on:

a. How many tasks in each of the five methods will be required for DT&E?

b. What is the average cost per task for the five kinds of DT&E verification tasks, based on past history?

c. How many items and interfaces will require qualification and acceptance?

d. What is the average cost per task for the five kinds of item and interface qualification and acceptance verification tasks, based on past history?

e. What is the cost of a preferred PMP activity, based on past history?

From this data an estimate can be formed, to which a margin should be attached for risk mitigation. As each task principal engineer crafts a plan and procedure, they must determine cost and schedule requirements and present them with their plan and procedure for review and approval, defend their estimates, and make any necessary adjustments to plan and procedure if cost is not approved at the level requested. When task work begins, the principal engineer must take responsibility for managing the accomplishment of the task in terms of cost, schedule, and accomplishment of planned work.

6.4 FEEDBACK INTO PRODUCT MODELS AND DESIGN

The results from all of the verification work provide valuable evidence that can be used in another powerful way. Throughout all of the development work, the engineering team will have used many models and simulation entities. Those tools will have produced certain results that were used as the basis for design solutions. Once the real product is available, its actual operation can be checked and the results compared with projections from the models and simulations. If the results are the same, we may say that we have verified our models. If they result in different answers, we have some problems that should be resolved and work accomplished to cause the same results. If differences are allowed to persist, not only can we not have confidence in the model base for the program, but serious risks of an unknown nature may exist in the delivered product. Once we have verified models, they may be used subsequently in a fairly economical fashion to study potential modifications and to resolve reported problems observed in system use. When the actual results match the analytical models, we can say the development process is characterized by engineering integrity.

Note in this case, where the models and actual performance differ, it is possible that the models are incorrect, or it is possible that the design is incorrect. Thus, before we apply the models to the development of the system, we should have first validated the models by ensuring that the logical and mathematical basis for the models is correct and in agreement with the requirements for the system. As part of this process, it is important to apply fundamental configuration management principles to the control of all of the product representations, including the models as well as the engineering drawings. All too often we apply careful configuration control on the design of the product, but fail to do so on the models that precede the release of the design.

6.5 TECHNICAL DATA ASSESSMENT

If the product includes operations and maintenance technical data, that data must be validated and verified. DoD applies the word *validation* to the first step in this work, meaning that the contractor demonstrates the tech data and corrects any errors identified. This work is accomplished by people intimately familiar with the product from an engineering perspective, but they may not be intimately familiar with the way the customer is going to apply the system to accomplish its mission. Subsequently, the technical data is verified by user personnel from a very different perspective by people who may have very recently completed related training on the system. Commonly, a different set of corrections will result from these two activities.

6.6 CONTRACTUAL CONSIDERATIONS

It is possible to include language in the contract that will make it very difficult for the contractor to be successful in reporting verification information and status to the customer. This condition will lead to bad relations between the two parties, unnecessary cost, and schedule slips. This language may be encouraged by past poor verification performance of contractors in general, because it is fairly easy for a contractor to make simple errors in reporting the evidence of compliance in such a way that it is very difficult for the contractor or customer to determine the degree of compliance actually achieved. At the same time, it is very easy for the contractor to truthfully make it perfectly clear what the status is at any point in time, but the author has seldom observed this condition achieved on a program because of a simple error.

Contractual language should encourage/require contractor submission for review of periodic status reports in terms of verification tasks planned and tasks completed and submission or availability of verification plans, procedures, and reports that give evidence of the quality of the work accomplished in those tasks. The content of these documents must be clear, especially the reports that must clearly provide linkage between the requirements covered in the related task and the location in the reports of the evidence of compliance observed during the task. Customer concerns for the content of these documents can be informally discussed between specialists on the two sides of the contract, and if a concern is sufficiently important the conflict can be made the subject of a technical interchange meeting.

Item qualification verification should be completed at a formal functional configuration audit where the customer reviews all or a selected subset of the many item verification reports and reaches a conclusion about the degree of design compliance with requirements. Alternatively, the customer may select some number of specific requirements and require the contractor to provide evidence of compliance in reports. Should there be a difference of opinion about the degree of compliance, then adjustments may have to be made, possibly involving a better contractor explanation, a design change, or retest witnessed by the customer. The final result should be full agreement that the design does or cannot affordably be made to comply with requirements.

6.7 VERIFICATION DOCUMENTATION APPROACH

Earlier editions of this book encouraged the development of integrated verification documentation such that each program would publish a single integrated verification plan, a single verification procedure document, and a single verification report document, each containing all of the documents of that type across the four verification classes. The author has since concluded that this was a case of overengineering the documentation, especially since he had never in several decades of experience observed anyone following this advice. So, this edition offers support for individual documents of each kind for each task, while offering a brief explanation of the integrated approach.

6.8 IN THE AGGREGATE

When an enterprise is developing a new product system in a program, the enterprise is really dealing with three systems interacting, and those responsible for each of them seldom realize the complexity of the construct with which they are dealing. A well-run enterprise that develops and delivers systems to customers should treat itself as a system, each of the programs it has in operation at any time as systems, and the product system that each program is developing as a system as well. Ideally, an enterprise would apply verification to all three. The correct agent for enterprise and program verification would probably be the quality department, verifying that the enterprise has published process documentation, that it was being faithfully and effectively applied on all programs, and that program plans and procedures were being followed in each case. These are matters that fit best in context with a book on system management in the author's view today, although a good case could be made for the whole verification story being under the quality umbrella. This book encourages that product system item qualification and system test and evaluation verification all be under the overall control of system engineers on a program, but that item acceptance be implemented by quality assurance on a production line.

The generic process employed on all programs should be designed based on a set of requirements contained in a world-class standard such as ISO 9001 for quality and system development in general, EIA 632 for system engineering, and IEEE/EIA 12207 or EIA J STD 016 for software. One approach to process verification is to evaluate the process versus a capability maturity model (CMM) such as the Carnegie Mellon Software Engineering Institute CMM for software or EIA 731 or ISO/IEC 15288 for system engineering. CMMI (integrated) from Carnegie Mellon is also available. These CMMs may be applied to the enterprise generic process and/or to one or more program processes going on within the enterprise.

On programs, the features of products developed should be compared with their requirements to determine if the design process produced designs conforming to those requirements. This activity, for the most part, must await the manufacture of product items that may be subjected to tests, but the planning work for this activity can start quite early. The time to start this planning work is during the development of the specifications for the product entities. That is the powerful message communicated by the V development sequence briefly discussed in Chapter 1. While we develop the requirements for an item, we should obligate ourselves to prepare the requirements for the verification process through which we will later prove design compliance with those requirements. The verification process planning and procedures (process design) should then be driven by these verification requirements.

Some people are comfortable with refraining from writing verification requirements in certain program situations, based on the belief that they do not contribute value commensurate with the cost of preparing them. But the value in doing the verification requirements closely coupled to the writing of the product requirements and written by the same person in each case is that if the product requirement is unverifiable, it will be detected in the difficulty of writing the verification requirement, hopefully triggering the solution of rewriting the product requirement such that it is verifiable. This becomes a matter of paying the price early in the program to avoid potentially tremendous costs later in the program, when the program enters item qualification.

For the author, the importance of this relative risk connection was vastly increased in his mind when he was asked by a legal firm to evaluate its client's performance in system engineering during an earlier program phase. At the time the program was in production and was $100 million overrun and 4 years behind schedule. The legal firm was concerned that, in a lawsuit involving the program, the customer might be able to claim that their client performed system engineering badly early in the program and that was the reason they were in such deep trouble. After a week of studying the program, the author had to inform the legal firm that their client had indeed done the system engineering work badly in specific ways that would have contributed to the cost and schedule situation. As a result, he was not asked to testify, of course, and very likely his report was never seen again. In this case, it was not discovered until item qualification how badly the specifications had been written. It was not clear what they were trying to prove in qualification testing. Acceptance testing was also a terrible mess. Much of the cost and schedule overrun occurred in item qualification and acceptance. Sadly, it is true that we often only find out when we enter qualification how badly we performed system engineering earlier in the program. This is far too late and unnecessarily so. If you wish to do requirements and verification work affordably, then do them well when scheduled.

The product use or implementation process should also be verified by the customer after they have taken delivery. Better yet, some customers like to have at least some of these characteristics verified prior to taking delivery. For example, it is possible to verify that it is possible to remove and replace items within a specified time period through controlled demonstrations, that a pilot can manage the resources of an aircraft efficiently while controlling the flight path through purposefully stressful demonstration flights or simulations, or that the product will provide service for the anticipated life through life testing. All of these verification actions do cost money and time, of course, so the customer has to balance their cost risk between development and operational phases.

On one program during which the author served as a field engineer in a seven-month operational evaluation aboard the USS Ranger in Tonkin Gulf, the program manager rightly conceived that all prior programs related to this product – an unmanned reconnaissance aircraft – had the same user, the Strategic Air Command, and new models and related support equipment had been integrated into continuing use of that system. In this new case, the company had to build a new set of support equipment and completely outfit a new unit at one time. The program manager sensed that we could conceivably have missed some important things in the design and set up a support equipment demonstration that the author led.

This demonstration, in combination with the system test and evaluation work, turned out to be of value. It was found that an engine inlet scoop that was thought to be essential in starting the jet engine on the launcher was blown completely off the launcher by the rocket blast of the launching rocket, resulting in the conclusion that it should be dropped from the design. Since the aircraft was to be recovered using a parachute, with the bird hoisted from sea surface to carrier deck by the SAR helicopter, the vehicle would have to be disassembled and all salt water removed between flights. This included a special engine decontamination tank in which fresh water could be passed through all exposed engine areas. It was found that the engine mounts for this tank had been located in the wrong places because the design engineer had used the wrong engine type as the basis for the design. There were several other minor problems also exposed. This activity was actually a part of the program verification process, but was thought of at the time as a special process due to the unusual case of a new customer with no historical experience with the system, except with the aerial target version of the basic airframe.

6.9 SUMMARY COMPARISON OF CLASSES

The next four chapters each cover one of the four verification classes. The reader should recognize a similar theme running through these four chapters, because the same basic methods apply across them all. There are some unique elements in each class, it is true, but in general all apply. Item acceptance differs because it must be integrated fairly tightly with item manufacture while PMP, item qualification, and system test and evaluation verification are more closely coordinated with the engineering work on a program. PMP differs to the extent that there will be thousands of cases in this class and each will generally involve a comparison of supplier data sheet content, many of these sheets unique in form, to the supplier with product requirements:

a. Define product requirements that in each case define essential characteristics that the design of the item covered by the specification must possess.

b. For each product requirement, define one or more verification requirements that define the corresponding characteristics of the verification work that shall be accomplished to determine the extent to which the design and/or manufacture of the entity complies with the product requirements.

c. Identify a level (item, child, or parent) of application and method (test, analysis, examination, demonstration, or special) to be applied relative to each verification requirement.

d. For each product entity, partition the complete set of requirements into task and method oriented subsets and assign a principal engineer for each of these tasks.

e. Assign each task principal engineer the responsibility to prepare a task plan and task procedure that will be subjected to a formal review and release and which the principal engineer will use in accomplishing the assigned task at the time assigned in the program schedule.

f. Assign a budget for accomplishing each verification task and identify any other resources that will be required.

g. Ensure that each task principal engineer accomplishes or causes to be accomplished their assigned task as planned and prepares a task report containing the evidence derived from the task pertaining to the question of the degree to which the product design and/or manufacture complies with the assigned requirements and contains a table that coordinates each requirement assigned to the task with the precise location of the evidence of compliance in the report.

h. Management formally reviews each verification report and approves the report after any directed changes have been accomplished, with the results being placed under configuration control such that they cannot be changed without subsequent formal review and approval.

i. Upon completion of item verification, subject the aggregate set of reports for a product entity to an audit to reach a formal conclusion about the extent to which the design and/or manufacture of the product complied with the requirements in the specification.

Of these nine activities, the first three are accomplished as part of the preparation of a specification, while activities "d" through "i" are accomplished as part of the item verification work for an item. Activity "d" involves the design of an item qualification or acceptance verification process in terms of identification of some number of tasks, each one of which has a responsible principal engineer assigned. Activities "e" through "h" coordinate with the tasks that must be accomplished in association with a specific item. The activity "i" audit is focused on a product or interface entity completing qualification or acceptance covered by a performance or detail specification. The audit input includes the item specification and the evidence of compliance of the design relative to the content of the specification that is contained in some number of task reports.

CHAPTER 7

Parts, Materials, and Processes Verification

7.1 CHAPTER OVERVIEW

This chapter explores effective ways of dealing with parts, materials, and processes (PMP) on a program. A preferred organizational structure for these activities is offered, as well as a way for programs to accomplish the related work. Traditional parts, materials, and processes work is commonly accomplished on programs using a preferred PMP list. Unfortunately, PMP verification too often is not really formally performed on programs. Parts, materials, and processes availability for manufacturing is often accomplished separately by personnel from different departments and a procurement department. What we intend to do in this chapter is consolidate and integrate these several activities within a functional department that can deploy a complete capability to programs. Appendix A9 offers a template for a program PMP plan.

7.2 PARTS, MATERIALS, AND PROCESSES ORGANIZATIONAL LINKAGE

The author would prefer an enterprise to have in its functional engineering department structure a Parts, Materials, and Processes department rather than have these components split into two or three separate departments. This department would create and maintain an enterprise preferred PMP list that all programs would be encouraged to employ as a starting point. The enterprise list would be evaluated by PMP personnel assigned to each new program and decisions reached about the current content relative to the new program. New PMP proven out in a new program could be considered for addition to the enterprise list as well as the program list. It is not intended that the PMP department should necessarily be completely self-contained, but rather that it provide

System Verification
http://dx.doi.org/10.1016/B978-0-12-804221-2.00007-3

the enterprise and its programs a coherent capability stretching across the three component parts. It is likely that some knowledge domains will better remain retained within other functional departments, where the knowledge is more tightly linked in program applications and applied to PMP needs as required.

7.3 PROGRAM PREFERRED PMP LIST

As shown in Figure 45, the enterprise list should be evaluated by new program personnel for general compliance with the needs of the program and compatibility with the program contract. If the enterprise list passes this test, it could become the initial program preferred PMP list. The PMP department would receive critical feedback from program PMP personnel based on their discussions with program design engineers, generally involving encouragement to update the list for new PMP. Any favorable conclusions by the PMP department would be approved and changes made. Some changes suggested might be accepted for inclusion on the program list but not accepted for the enterprise list. A third possibility would be that the PMP department could reject listing on either the enterprise or program list, possibly leading to a higher tier organizational discussion.

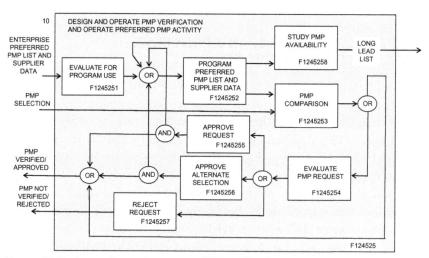

Figure 45 Design and operate program PMP verification.

This evaluation must include a study of the supplier data for consistency with program requirements to the extent that it can be determined at the time. Any changes found necessary should be made before the program starts design work that will cause design engineers to have to make PMP choices. Over the run of the program, particularly during preliminary and detail design, it is likely that program design engineers will encounter difficulties with the program list omitting new PMP that they find favorable relative to what they had been considering using in the design, so throughout the design phase of the program there will be a need for interchange between program PMP personnel and design engineers. Any of these changes should be extended to PMP department personnel for consideration for inclusion on the enterprise PMP list. These kinds of conversations should be going on between program PMP personnel and department PMP personnel as a way of continually updating the enterprise list always employed as the initial program PMP list.

As product and interface entities enter qualification, it may develop that PMP failures during testing could lead to the conclusion that PMP were selected from the preferred PMP list that will not survive in the environment within which the product is required to function. Alternatively, it could be determined that the testing performed stressed the product in excess of the requirements in the specifications. In either case, program adjustments will have to be made, either in the choice of a PMP, a design change, or both.

Over time the enterprise PMP list may include two or more alternative PMP, from which the design engineers could select. Different PMP could be used for the same purpose, which design engineers might find advantageous in the form of cost, performance, weight, or other factors in context with other program needs, so it is not necessary to ensure that there is only a single PMP listed for a specific purpose. PMP that can no longer be purchased should of course be dropped, no matter their other benefits.

It is understood that an enterprise PMP list will inevitably be a moving target over time, because the availability of the listed elements will commonly have a limit in how long their source, from a business perspective, remains interested in providing them in their product list. Therefore, each new program offers the enterprise an opportunity to refresh the list.

7.4 PMP VERIFICATION

The very act of operating a preferred PMP activity on a program provides an effective PMP verification service for the program. The enterprise and its programs are aware of supplier PMP capabilities and they can easily acquire this information from suppliers. The process of design engineers selecting PMP from the program preferred PMP list only requires that these design engineers assure themselves that the PMP selected comply with program requirements captured in the lower tier item performance specifications. An enterprise could establish policy for its programs that design engineers have this responsibility as part of their design task. Alternatively, an enterprise could operate a PMP verification activity where supplier data for PMP called out on engineering drawings must be checked for compatibility with the requirements of the performance specification for the item in which it appears. Clearly, the latter will cost more and probably extend the schedule demand to some extent over the former. Either scenario could be well implemented on a program, or be implemented badly with no clear evidence of potential problems, so in either case the work must be managed well. The author would load up the design engineer with this responsibility, since he or she will be as close as one can get to the knowledge needed to make the choices for a particular design.

The PMP verification process appears in Figure 45 as function F1245253, where someone must verify that PMP selected by the design engineer compares favorably with supplier data. The question is, who makes this comparison in a particular enterprise and program? This can often become a heated battle between functional department managers, best left to the cooler heads they work for, if they exist.

7.5 PARTS AND MATERIALS COORDINATION WITH AVAILABILITY SEARCH

Early in a program it is helpful to know if any parts and materials being considered for inclusion in the design of the system will become unavailable for procurement during the program manufacturing run, and customers may be interested in information about maintenance parts they will have access to subsequent to entry of a new system into their service. Function F1245258 is included on Figure 45 for this purpose. Companies exist that will conduct a survey of suppliers to find out what their plans are for particular parts and materials and offer this service to companies for a fee, either for a specific list of parts and materials for a particular period of time or as a long-term service for some particular range of parts and materials related to a client's customer base and system type.

One solution to parts and materials availability over the life of a program is a lifetime buy, where the program must determine how many or how much of the part or material will be required over the life of the program in terms of qualification needs, initial production, and the potential for follow-on production, perhaps with customer cooperation in taking some of the cost risk for logistics support for spares, repair parts, and materials. Another alternative is to plan for upgrade of the design, where some items that parts and materials suppliers plan to delete are replaced in a timely way relative to supplier plans. Clearly, there is considerable cost risk in any of these life issues, often leading to attempts to sell off parts and materials purchased in excess of actual needs on a program.

7.6 LONG LEAD LIST

During preliminary design activity, as design engineers identify parts that will be needed to manufacture items that will have to be subjected to item qualification, it will be necessary for parts, materials, and processes people to determine any parts and materials that will be hard to acquire in the quantity required to fabricate those items needed to support item qualification and place them on a recommended long lead list, which will have to be reviewed and approved by the program manager. As early as possible, this list should be submitted to the customer program manager for approval. There may be considerable discussion about the items on this list because of the risk that some of the items on the list will be dropped from the list as the design matures. Any items dropped from the list after they have been purchased will have to be disposed of in some way.

Figure 46 shows the logic supporting the long lead concept. If a program waits until critical design review (CDR) to request authorization to purchase long lead items, it is likely that they cannot be acquired in time to support manufacture of the items needed for qualification testing, causing a corresponding schedule slip of subsequent program tasks. At the time that these items must be manufactured, there will be no production line, of course, so the items will have to be manufactured in an engineering shop by technicians and engineers.

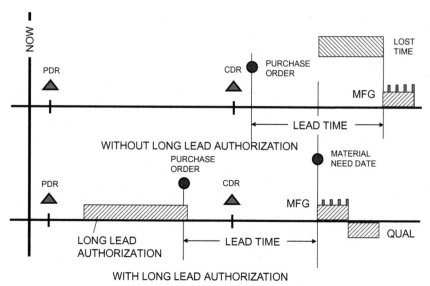

Figure 46 Long lead item logic.

7.7 SOFTWARE PMP

In his career in industry, the author has found very little interest in PMP thinking in terms of software development. A lot of the thinking and work that hardware people put into PMP could benefit software development. Where a program intends to purchase a software "part" to include in a portion of the code for the program, the program would do well to ensure that the part will function as claimed, that its configuration will remain stable for the duration of the program, that there is a solution to the continued legal availability of the part in the event of the termination of the supplier company during the run of the program, and if then software part is needed early in the qualification process that it can be included in the program long lead list.

CHAPTER 8

Item and Interface Qualification Verification

8.1 OVERVIEW OF THIS CHAPTER

This chapter covers the second of the four verification classes, called *item and interface qualification* verification. Item and interface qualification is applied to product and interface entities below the system level and above the part, material, and process level. The purpose is to determine the degree to which the design of an entity complies with the requirements contained in the item or interface performance specification. The item or interface performance specification defines the problem confronting the design team. That document should clearly define the essential characteristics that the product entity must possess. The design of the entity is intended to provide a solution to that problem. It then remains for us to prove that the design does in fact offer a complete solution to the problem, and that is what it is hoped item and interface qualification will do. The qualification process is intended to qualify an entity for use in a particular application defined in its specification.

As noted earlier, we should really not look at the purpose of verification as proving that a design complies with its requirements. This can color our attitude, to give preference to the desired outcome. The correct attitude about verification is that it is intended to determine the relationship between the requirements in the specification and the characteristics of the item being verified. Those involved in verification must be mindful of the need to preserve the integrity of the process and those involved in accomplishing it.

While it would be possible to simply attempt to take off on a test flight in a newly designed aircraft, composed of hundreds of component parts, as a way to verify that the whole airplane complied with its requirements, customers and those who develop such products, to say nothing of the flight test crew, have discovered largely through the experience called "hard knocks" that it is better to first determine the qualification of components for an application before assembling and attempting to operate the whole end item. There is clearly both an economic and safety concern involved.

System Verification
http://dx.doi.org/10.1016/B978-0-12-804221-2.00008-5

In this chapter we will introduce the concept of promotion or demotion of verification work for an item, and moving directly to flight test for a new aircraft would be the ultimate folly of applying the promotion technique for all aircraft parts, materials, processes, items, and interfaces described and defined in hundreds or even thousands of specifications. Risk, our ability to measure it, and our ability to adhere to what effective risk managers tell us about its reduction are important career extenders for those who would manage verification work.

8.2 OVERALL ITEM QUALIFICATION PROCESS

8.2.1 Where Does the Qualification Verification Work Begin?

We enter item and interface qualification verification work after completion of the item or interface performance specification, including its verification section and a verification traceability matrix listing all of the verification strings or records that apply to the item. Included in the matrix will be requirements that we will partition into subsets, each one of which applies to one of the five methods as well as many titles-only paragraphs that will coordinate with a sixth method called "none." For a substantial program, the number of performance specifications may be very many, but we can treat them all in exactly the same way with respect to qualification verification.

For each of these specifications, it is recommended that the work of preparing the specification be completed by a small group of engineers, each of whom would have written one or more product requirements, derived through use of a modeling application, for inclusion in the specification. The engineer writing each product requirement must then identify the recommended method of verification and prepare one or more corresponding verification requirements. As a result, an included verification traceability matrix would come into being as the specification content is prepared.

It is entirely likely that everyone who contributed to the specification would not have the same experience and background in doing this work, so the result may be somewhat uneven. A system engineer, hopefully with experience in verification process design from prior program work, would have been assigned the overall responsibility for each specification or some group of specifications and this engineer should review the evolving content of the specification and verification traceability matrix and discuss any criticism of choices made by contributing engineers, so as to provide an integration and optimization stress to the development of the requirements content and the verification plan inputs derived from the specification.

So, to partition all requirements derivation and specification publishing work to the left of some date and all of the verification work to the right of that date is not quite accurate. Commonly, the requirements work tails off as the verification work starts to pick up for a given entity. In this book we place the formal transition between requirements and verification work arbitrarily with the completion, review, approval, and release of the specification that includes the work of writing the verification requirements and development of the traceability matrix included in the specification that includes identification of the method of verification and level of accomplishment for each product requirement. It is not necessarily true that no verification planning work should be accomplished prior to this event, but any work preceding it may have to be revised if the specification content further evolves prior to review, approval, and release of the specification.

With the specification complete, we can begin the difficult verification process design chore, which involves determining how to package the many requirements in the specification into some finite number of verification tasks of the six (including none) methods. One alternative is to assign each requirement to a separate verification task. This alternative is, on its face, silly as well as expensive if actually implemented. The other extreme is to assign them all to one task of one of the five methods, which is also not sensible. Some guidance on how to proceed in order to realize the condition of best evidence at affordable cost might be helpful in this matter, and is offered in this chapter.

8.2.2 Good Works and Integrity Again and Again

Some people believe that the purpose of verification is to prove the design satisfies the requirements; this is not the case, except, perhaps, in the minds of some program managers. The purpose, as it applies to item qualification, is to determine the relationship between the design requirements for an item in its performance specification and the characteristics of the design as represented by a product entity manufactured as closely as possible with the plan for full-scale production. We are interested in the degree of design compliance with those requirements. Verification is a process of comparison of an entity of interest with a standard of excellence followed by a conclusion about the degree of compliance of the entity of interest relative to the standard, based on evidence achieved through the application of a valid method of inspection of the entity of interest. In the case of item qualification, the standard of excellence is the content of the item performance specification and the entity of interest is the design of the entity that is the subject of that specification.

A high degree of integrity is required on the part of the engineers who perform this work, because the engineers doing the verification work will often be criticized by persons in supervisory or management positions if the verification work they accomplish "fails to show compliance" with the specification content. Engineers performing verification work must remain focused on the *truth* regarding that comparison, but not everyone on a program may have that same dedication. Some may be more interested in budget and schedule than the truth in this comparison. It is one of the great truths of systems engineering that an enterprise is badly served by engineers and managers who fail the integrity test offered by verification work.

With a good specification, verification plan, and verification procedure, followed by the verification work performed by engineers with the prerequisite skill and knowledge, if the design fails to comply with its requirements, then a bad product design is not an uncommon cause. Persons doing this work must have the integrity to speak the truth, however unpopular it may make them.

Commonly life is not as simple as this, however, in that the program will have generated at least some bad specifications that include badly worded product requirements, missing requirements, and unnecessary requirements coordinated with verification requirements that were a challenge to prepare. The design work may have been badly influenced by unclear specification content. Combine this beginning with the possibility that one or more verification task principal engineers may have prepared a poorly conceived plan and procedure, and one begins to understand why so many programs encounter a great deal of trouble when moving into item qualification verification. The problem becomes even more acute, since there is not much cost and schedule margin remaining on the program manager's books by the time the program arrives in verification.

However, this story does not always have a bad ending. The way to encourage a good ending is to prepare good specification content that was crafted based on a sound modeling approach, which the engineers doing the work have become experienced in applying; extend this good beginning with well-stated verification requirements, coordinated with the best methodological choices, and complete the preparation with the development of a set of effective item qualification verification process designs expressed in an affordable set of verification tasks, expressed in each case in a good item verification task plan and procedure using the techniques encouraged in this book.

But even with all of this present during implementation of the verification work, a program can still fail to succeed through poor reporting of the results of the verification tasks. Beginning with the assignment of tasks to principal engineers, each task principal engineer should be given a list of the particular requirements for which he or she must produce evidence of the degree of compliance achieved by the design. That list must be included in the task report summarizing the outcome and telling *precisely* where in the report the degree of compliance is stated. The report must state the evidence very clearly, such that whoever reads or observes it will easily reach the same conclusion. To achieve this outcome it is not enough, of course, to have good practices for even good people will not always do the right thing. It is necessary that programs be very well managed during this verification work to realize best results. If followed, the conditions described in this paragraph would have caused many programs that encountered major difficulties in verification to instead sail through the experience, with many compliments from their customer.

8.2.3 Process Flow Diagram Expansion

Figure 47 offers a flow diagram for the complete item and interface qualification process, showing three different kinds of qualification patterns. The normal case is referred to in this book as the high rate production pattern. In this case there are enough articles of each kind produced that one or more articles can be manufactured especially to support qualification, with the understanding that they will be stressed in qualification such that they will not satisfy production unit reliability or life requirements and cannot be delivered to the customer. Two other program patterns are available: (1) one-of-a-kind, and (2) low-rate, high-dollar.

Note that Figure 47 refers to an item M intended to be one of many product entities or interface entities that must pass through qualification. So, this task related to item M will have to be repeated for each entity involved in the system being created. Each of these item qualification verification actions will be accomplished in accordance with a schedule that will have to be adjusted from time to time based on opportunities and adverse outcomes of earlier work. It may require a week or two to complete one item M qualification task N, five months to complete all of the item M tasks, and the complete qualification work span may stretch over a period of two years on a large program. The qualification of item M may require accomplishment of one task or 113 tasks completed in some combination of the five methods. For a real program it might be possible to prepare a flow

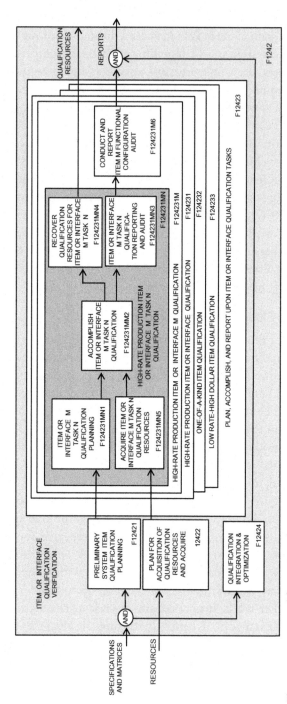

Figure 47 Item qualification verification.

diagram reflecting each and every separate task, but in this book we will employ a generic approach, recognizing that it will falsely imply a degree of simplicity that is not warranted. Figure 47 appears to reveal only a single task N being performed, but the reality is that there are many tasks N arranged in some series-parallel pattern of work for each item M124231M. In some cases an actual product entity will be required to test, demonstrate, or examine in accordance with a procedure, while in other cases the task will call for an analysis of related data or modeling and simulation work, requiring no access to a physical product at all.

Figure 48 offers a view of function F1242 focused on a one-of-a-kind program involving the production of only a single article. The three lead-in functions are repeated from Figure 47 with just the one-of-a-kind function expanded. The only difference between the tasks shown in Figures 47 and 48 is that the one-of-a-kind program will require higher design margins and lower verification stress, because the production article must be delivered with a full life expectancy. This kind of program may also encourage more use of the analysis, examination, and modeling and simulation (special) methods and fewer test and demonstration tasks than might be employed on a high rate production program. The one-of-a-kind case will often also have to make allowance for a need for some refurbishment of the article subjected to verification to account for the chance of product life or reliability loss.

In the low-rate, high-dollar case shown in Figure 49, a small number of very high value units will be produced and delivered, making it difficult to justify production of units that are dedicated to item qualification where stresses will commonly damage the units. A compromise is often employed where the first article is passed through a subdued qualification, subjected to selective refurbishment, and delivered last, acting as a spare with the hope it will never have to be used for an operational mission. The only application of this case the author has observed happened on a program in great difficulty, but that failure did not cause the author to conclude that the methods employed should never be employed. That program, badly managed from the start, developed poor specifications, a flawed design, and bad verification planning. The tremendous cost overrun and schedule delay, which resulted in a legal battle between the customer and contractor, could have been avoided.

No matter the verification process employed, every product entity delivered to the customer must be capable of providing the customer with full life service and this is the primary problem with both the low-rate, high-dollar and one-of-a-kind variations of verification. We are unable to apply the

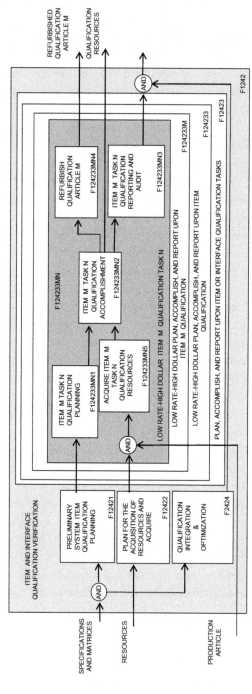

Figure 48 Item and interface one-of-a-kind production qualification verification.

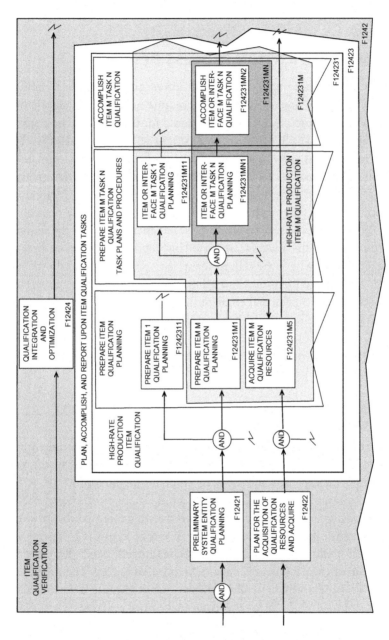

Figure 49 Item and interface low-rate, high-dollar qualification.

kinds of stresses to an item in these cases that we can apply in a high-rate production program, because those items will never be delivered to a customer.

8.3 ITEM QUALIFICATION PROCESS DESIGN

Our item and interface qualification verification process calls for identification of item and interface entities that will require qualification, all of which will have a performance specification prepared for them. All of the requirements in Section 3 of each of those specifications will be coupled with a verification requirement in Section 4 and the pair will be listed in a verification traceability matrix in the specification. Engineers will decide what method should be applied in verifying each requirement and at what hierarchical level (item, child, or parent) the related verification work should be accomplished. At this point the real verification work can begin. All of the requirements from each specification are imported into a program-wide verification compliance matrix where the requirements are sorted by method into specific tasks. Each task is identified by a unique task identification as covered in Paragraph 6.3.2 and a principal engineer is assigned to each task. This principal engineer is then responsible for preparing a plan and procedure, arranging for availability of task resources needed, accomplishing the verification work in accordance with the plan and procedure within available budget and schedule limits, and preparing a report of the results.

8.3.1 Qualification Task Assignments and Integration Opportunities

The input to the process of determining the task in which a requirement will be verified is the content of the verification traceability matrix in the performance specification for the item or interface in question. All of the requirements assigned to a particular task should share a common verification method. If it appears to the responsible task engineer that the information in the verification traceability matrix is wrong regarding the method identified for requirements mapped to the engineer's task, the principal engineer pursues a specification change before making any change to task planning. The requirements collected into one task must share the same method but should also call for the same supporting resources, such as test equipment.

The period when all of the requirements are being collected into tasks offers a program a great opportunity for integration opportunities, so should be followed closely by one or more system engineers. It is possible that requirements changes, method changes, design changes, and other possibilities could not only result in a verification cost reduction but also an improvement in the product delivered to the customer. It is often difficult to reach a conclusion about these situations because it is necessary to compare the benefits of a change with the adverse effects of that change. If a change results in a savings of $43,000 in qualification verification, but will cost $150,000 to implement, then it is probably not a good program choice. If it would also result in significant manufacturing and acceptance verification cost reductions, it might prove to be advantageous. One does qualification once on a program, but acceptance is accomplished on every article manufactured.

Engineers responsible for design of factory and field test equipment may be making design decisions during this period also and it is possible that some of this equipment could be useful in qualification, or that qualification work could be helpful in guiding equipment design decisions. Persons responsible for field technical data may also show interest in the evolution of test procedures employed in qualification verification.

8.3.2 Developing the Task Plans and Procedures

In preparing for item qualification, we must first plan the work, extending our understanding of the verification requirements for a given item into a definition of some number of tasks of the five methods. Each of these tasks is then planned in detail, a procedure is prepared, and persons or teams made responsible for accomplishing the work. The planning data can be collected into a single *integrated verification plan* (IVP) or, more commonly, a set of individual verification task plan and procedure documents, which this book considers the norm. Previous editions of this book encouraged integrated program-wide verification documentation, but the author has come to accept that this is a case of overengineering a simple problem best dealt with as a collection of tasks, each independently planned but subjected to integration and optimization across the planning work. A single document containing all program verification plans, procedures, and reports is simply excessively difficult to produce, the resultant document too extensive in content, and the resulting program too difficult to manage. Also, the author has not found anyone interested in employing the integrated document approach in the 20 years since it was first included in an earlier edition of this book.

Figure 50 represents a focus on and decomposition of the front end of the high-rate production item qualification process (F14231) shown in Figures 38 and 47, identifying several significant subtasks. Figures 47, 48, and 49 offer the three item qualification patterns, but this chapter will describe primarily the high-rate-production pattern. On a high-rate-production program, sufficient articles will be produced to justify the cost of producing some small number of qualification articles, which will then become unacceptable for delivery to the customer as a function of the qualification work, which reduces item life expectancy and reliability while inspecting item compliance relative to performance specification content. As we will see, these expended articles will often have residual value in program verification and training activities, but they should not be delivered as possessing a full life. Only item M tasks 1 and a generic task N are referred to in Figure 50, but as noted earlier most programs will have to deal with many product entities. Also, any task N may have to be implemented in one of five different methods not detailed on the figure.

A program does well to prepare a system verification plan at the F124 level and qualification plan component at the F1242 level early in the program, or this content can be included in the program integrated master plan (IMP) on a small program. The plan defines data that must be tracked over the whole process that will be best contained in the context of a relational database, as discussed in an earlier chapter. The three principal manifestations of this work are the verification compliance matrix, referred to by some as the verification cross-reference matrix (VCRM), the verification task matrix, and a verification item matrix. The former lists every requirement in all of the program specifications, correlates the details of the verification process, and provides the principal verification process design tool. The second is the principal process management tool, defining all of the work at the verification task level that coordinates with budgeting and scheduling management interests. The third matrix maintains a record of program progress relative to the items that require verification useful in planning for item-oriented verification audits. This data must be maintained throughout the verification process, as planned work is accomplished to provide management with the information needed to support sound verification management. All three matrices can be supported from a single relational database, ideally the same database that supports requirements analysis and specification publishing.

Figure 50 opens with a preliminary system item qualification planning task (F12421), where those responsible for the overall qualification activity

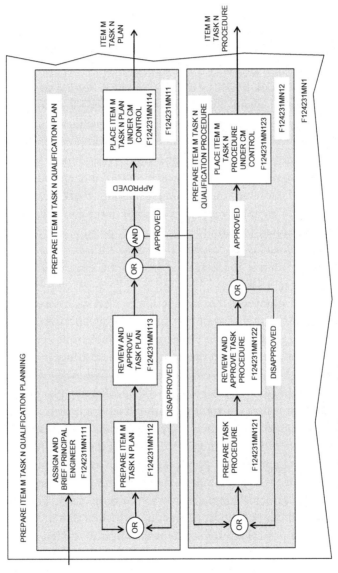

Figure 50 Item qualification verification front-end planning.

design the intended qualification work pattern and make personnel assignments. The verification compliance matrix input includes all of the item product requirements from Section 3 of the item and interface performance specifications, all of the verification requirements from Section 4 of these specifications, entries for the method to be applied and level at which the verification action will occur in each case. This data all comes from the work that was completed to prepare the item performance specifications and their included verification traceability matrices.

The work under task F12421 can be as simple as making sure that every requirement is included and all entries have been completed for an item of interest. It would be helpful for a person in overall qualification verification responsibility to prepare a planning statement that those involved in more detailed parts of the plan could use as guidance. One or more system engineers should be involved in applying integration and optimization skills (F12424) to the evolving qualification planning work. Early in task F12424, this should include a cursory evaluation of the evolving set of qualification tasks identified and consideration of any possible alternative sets of tasks that might offer a lower-cost effective result. One of our overall objectives should be to derive the best evidence of the degree of compliance attained that we can afford.

The reader will observe that from function F12421 we have to spread our interest to all of the items that must be subjected to qualification, and each of them must undergo essentially the same process, which from a planning perspective includes allocating all of the item verification strings in the compliance matrix to tasks, each focused on one of the five methods. Figure 49 includes function F12421M1 to cover general qualification planning activity for an item M that was inadvertently omitted from Figures 47, 48, and 49.

A principal engineer is assigned to every task and preliminary cost and schedule figures are estimated, subject to review as the individual task plans and procedures mature. At this point, we become aware for a given item how many tasks of each of the five methods will be needed and who shall be assigned as the principal engineer for each of those tasks. These insights provide us with an item qualification design, and when all items progress to this point we will possess a program item qualification design with which management can begin.

We will need to identify each product entity and interface that will require qualification verification and every verification task for each of these items in function F12421. If an overall program qualification plan has not been prepared previously or included in some other program plan, work should not continue until that is complete. On a large program with many items needing qualification, someone should be identified as the principal engineer for each item at some level or levels of indenture. This could be someone from the responsible design team for qualification, but we should recognize that these teams will very likely be disbanded before item acceptance is complete. In this book we will take the position that a system engineer should assume the lead role for the system test and evaluation and item qualification program, but that a manufacturing engineer or quality assurance person on the system team could take on the responsibilities for acceptance verification technical leadership. It is in function F124231M1 that the item M principal engineers partition all of the verification strings identified in the verification traceability matrix contained in the item performance specification into specific verification tasks, each of one of the five methods.

For every task, we must identify a principal engineer who will take on responsibility for the task. On Figure 50 it is in function F124231MN1 that the task principal engineer prepares the task plan and the task procedure for item M task N. The plan will identify all of the resources needed to accomplish the task and the task principal engineer must follow through to insist on the program acquiring them. Most qualification tasks will require some resources, however simple, and function F124231M5 of Figure 50 is the beginning of this process carried on by task principals in function F124231MN5 in Figure 47. For a given task such as Item M Task N, we will have to wait until the plan has been prepared, but in function F124231M5 we can begin a list for item M tasks, updating it as the plans and procedures become mature. Some of these resources will require manufacture of an article that will be tested as the means of developing evidence of requirements compliance. Generally, most of the items that will have to be manufactured to support qualification will have been determined during early program planning work in order to determine a good budget estimate, but this list will likely require update.

Perhaps the most important activity in the whole verification process, which is so often neglected, takes place in function F124231MN111 on Figure 51. Each task principal engineer must be presented with a list of the verification strings from the compliance matrix that have been allocated to the principal engineer's task. The principal engineer must be informed at that time that he or she must include this table in his or her plan and procedure, with traceability to the specific plan and procedure content that will develop the evidence of compliance for each requirement. Also the task report will have to include the table, along with traceability, that states

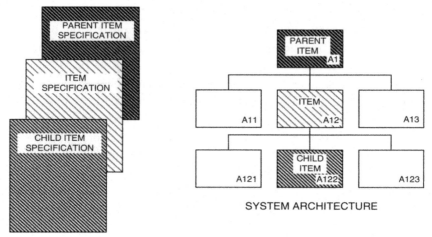

Figure 51 Item M Task N plan and procedure development.

for each string specifically where in the report the evidence of compliance is contained. Later in this chapter, when discussing the reporting, review, and approval process, we will pick up on this point, emphasizing that the report must be reviewed for this information and not approved unless the evidence is convincing and easily found. Task principal engineers will not receive this news happily and many will not comply willingly, but the management effort it takes to cause success in this matter will separate the affordable and successful verification work from the failed.

Figure 51 expands on the item M task N planning and procedure preparation activities. Clearly there will be many of these activities on a program of any substance. If, for example, a program will entail development of 52 items and each will require on the average 30 tasks, then the program will require that 1,560 (52 items × 30 tasks) plans and 1,560 procedures be developed. These numbers may be enough to encourage development of plans that also contain the procedure dividing the number of these documents in half and that could certainly be done. This motivation would be stronger when it is realized how simple some of these documents might be, involving a single page for some simple verification tasks. Another alternative would be to prepare a single plan and procedure for qualification of an item that integrates the plans and procedures for all tasks to be applied. The author will leave that decision to the reader and continue with a rule of separate documentation, while recognizing the possible cost appeal of joint plans and procedures.

A formal plan is a prepared method for accomplishing specific steps toward a specific end result. The objective of the qualification verification plan is to clearly identify a course of action that can be extended into a procedure that can be accomplished within available program cost and schedule goals, resulting in development of convincing evidence of the degree of item design compliance with a specific set of requirements contained in the performance specification for a particular item. This plan must identify instructions for preparation and for accomplishing the task in terms of the supporting resources, their condition, arrangement, and relationship; contain a clearly identified set of steps that must be accomplished in a particular order, recognizing that a procedure will contain the detailed stimulus-response relationships needed to actually produce final results; identify what information must be collected and how it can be determined whether that information is acceptable or not; and state who is responsible for leading and accomplishing the task.

If the task must develop compliance evidence for multiple requirements, the order of attack on those requirements will often be important, so care should be taken while forming the design of the plan to find a best sequence. Make an effort to determine a best sequence but consider alternatives and evaluate the effects. In some cases the results from verification of one requirement can be used to reduce the effort required to verify another.

Given an approved plan, or great confidence that the plan in work will be approved, a procedure can be prepared to implement the plan in terms of detailed steps for accomplishing it. In some cases, the steps will be very uncomplicated, possibly even involving a single sentence for an analysis task, but in other cases, often involving test or demonstration, the procedure can involve many steps and even a cyclical behavior on the part of the person, equipment, or software performing the work, with different responses determined by prior responses. In some cases, a probabilistic work pattern may be required where mean and standard deviation values are needed.

Various organizations have prepared many applicable documents to guide verification work, especially involving test, but not limited to test. Many customers will insist on some of these applicable documents being respected in verification work accomplished in development of their system. Even if this is not the case, many developers have found these standards to be effective. In other cases, developers have developed internal guidance documents that capture what has been found to be effective on its product line. In some cases, these internal documents are simply edited versions of more formal and general applicable documents.

The review of each task plan and procedure is critically important if you wish to be successful in implementing an affordable verification process, for we must review the documents to ensure that the principal engineer has included in the plan and procedure the necessary content to verify compliance with the requirements assigned to the task. Someone should read each plan and procedure and reach a conclusion that following it will produce insight into the degree of compliance the design achieved, relative to the requirements assigned. If these documents fail to do so, then they should be disapproved and subjected to change. It should also be determined that each plan includes the matrix for the task that lists the requirements to be verified.

It could happen that every task plan includes every string allocated to the task, but that one or more strings were still missed for the item. Therefore, someone should ensure that, as each plan is reviewed, the strings in the compliance matrix are checked off. As the last task plan is reviewed, we should find that all of the item strings in the compliance matrix are covered.

8.3.3 Level of Task Accomplishment

Item acceptance verification should be accomplished focused on the physical manufacturing of product entities. Item qualification verification, on the other hand, should be accomplished within a framework permitting a wider range of possibilities. It is possible to force all qualification planning to be accomplished closely tied to the content of each item specification, but a broader view will commonly result in acceptable item qualification verification evidence at an affordable cost. Figure 52 illustrates the possible product entity relationships and their correlation with specifications.

Most often, the best compliance evidence will be obtained through a verification task accomplished in association with the product entity the task is focused on. But there are cases where best evidence will be realized not at the item level, but at the parent or child item level. In order to permit this action, the program should permit promotion or demotion of item qualification verification. Promotion can be a useful part of an effective qualification process, but early in this chapter we cautioned against excessive promotion of verification action.

Figure 53 graphically reveals all of the verification task possibilities for the requirements in an item specification. First, the requirements in the specification may be verified through one of five methods (test, analysis, demonstration, examination, or special). For any given method, item requirements could be verified in association with tasks accomplished at the level corresponding to the item specification. Alternatively, they could be demoted to

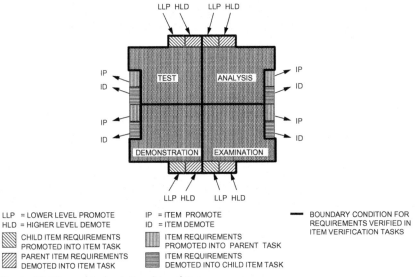

Figure 52 Hierarchical relationship of items.

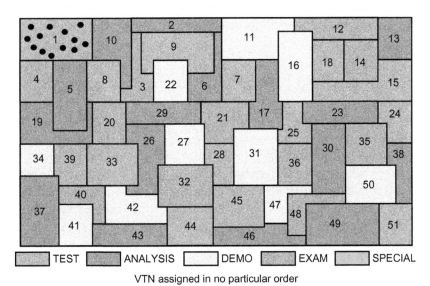

Figure 53 Requirement verification promotion and demotion.

tasks related to a child item specification (ID), of which there may be several, or promoted to tasks related to the item parent specification (IP). Also, a requirement from a child specification could be promoted to be verified in a task associated with its parent specification (LLP) or a requirement from a parent specification could be demoted to be verified in a task associated with a child specification (HLD). Thus, the requirements to be verified in association with tasks related to a particular item would fall into a set bounded by the heavy border in Figure 53 rather than the simple boundary formed by the item specification. Granted, this policy will result in a more complicated item qualification process to manage, but it will provide for the most affordable process possible if well managed.

It is possible that requirements in specifications for end items that next assemble into the system could be promoted into system test and evaluation level tasks, or that requirements appearing in the system specification could be demoted into a child specification, thus realizing a cross-class integration situation. Promotion of end item requirements to system test and evaluation tasks should be avoided because it can result in an unresolvable, circular verification logical problem. The customer may require that all item qualification verification tasks be complete and subjected to audit before system test and evaluation can begin. But, if item qualification is depending on one or more requirements to be verified through a system task, then this program requirement can never be satisfied.

8.3.4 Qualification Verification Process Design Example

The principal problem in verification is the design of the verification process. An overview of this work was offered in Paragraph 8.2. Now we wish to focus on the transform between the requirements appearing in a particular performance specification and the set of verification tasks that collectively will expose us to the evidence of compliance of the design with the content of the specification.

An example of collecting several requirements into one task can be observed when assigning reliability, availability, and maintainability (RAM) for all items in a system into one RAM analysis verification task, reported upon at the system level. All of this work could be accomplished by a single engineer on a small system, or by a small army of engineers, each focusing on one item for one of the three pieces on a very large system. The work could be separated into three separate tasks, or it could all hold together as a single task, be planned in a single plan, be accomplished in accordance with a single procedure, and be reported upon in a single report, recognizing that for a large system these documents would likely be in the form of multivolume sets. This would essentially be a case of promoting all lower-tier RAM requirements to the system level. In this example, we will not deal with promotion or demotion in the interest of simplicity, discovering that simple cases are difficult enough to get our arms around.

For each of the requirements within a specification, we are interested in the method by which we will pursue evidence of compliance (test, analysis, examination, demonstration, or special). We are also interested in the level at which we will pursue verification (item, parent, or child). Most often the choice will be item level, but there are cases where we should permit promotion to a parent entity or demotion to one or more child items. This can make the process design more complex but it can also result in a more affordable process. These decisions should be included in the verification traceability matrix included in the specification and are therefore inputs to the verification process design.

8.3.4.1 The Size of the Verification Process Design Chore

Before discussing the process of transforming sets of verification strings into verification tasks, let us first consider what the whole picture would look like for the verification tasks for one representative performance specification. Figure 54 illustrates a Venn diagram of all of the product requirements in Section 3 of a specification, where they have all been mapped to a verification task number (VTN), all of which are numbered VTN 1 to VTN 51. The paragraphs requiring no verification have been ignored in this figure.

A total of 13 requirements have been mapped to a test task (colored blue in the figure) assigned verification task number 1 and that task, like all of the

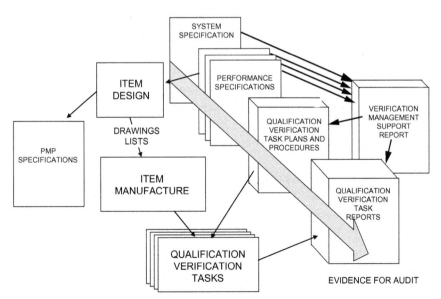

Figure 54 Verification string transform to tasks example.

others, will need a plan and procedure prepared that, when implemented, will produce convincing evidence of the degree of compliance of the product with the 13 requirements assigned to VTN 1. The same is true for the requirements mapped to the other 50 tasks for this item qualification verification work. Keep in mind that Figure 54 is but one example of how requirements might have been assigned to verification tasks. The possibilities are many.

Other than VTN 1, Figure 54 does not disclose how many requirements were assigned to the 51 tasks for the item. Table 2 shows an example of how the requirements could have been distributed between the 51 verification tasks in a four-column table listing the 51 VTN, and telling in each case the method that will be applied and the number of requirements (REQ) that were assigned to the task. For any one item and its specification, there are, of course, many possible outcomes of this effort and there is not necessarily only one right answer to the problem. People with experience and good engineering sense, motivated by management to devise a plan that will produce the best evidence of compliance consistent with available budget, will commonly come up with an acceptable proposed plan that can be reviewed, exposed to critical comment, and after some adjustment perhaps, be approved. Other persons doing this work might come up with a significantly different set of conclusions that might be every bit as effective.

As the reader can see, by adding up the number of requirements (REQ) for each task in Table 2, this specification includes a total of 147 requirements needing verification. There might be 23 other paragraphs in Section 3 that will not require verification of any kind. Table 3 extracts the number of tasks and requirements distributed in this example to the five methods identified in Table 2. The author does not suggest that either Figure 52 or Tables 2 and 3 offer a useful way to actually build a verification plan. Rather, these views of a single specification example are discussed only to expose the reader to the number of entities that must be considered while doing this work.

We have been discussing only a single specification, and on a large program we may very well have to deal with over 100 product entities, other than the system specification, with both a performance and detail specification for each item. If each performance specification (which includes the system specification) contained 150 requirements needing verification on average and each detail specification 95, then that would result in 15,150 requirements (101×150) in 100 item performance specifications and the system specification, and 9,500 requirements (100×95) in the 100 detail

Table 2 Sample Requirements Distribution by Task and Method

VTN	Method	REQ	VTN	Method	REQ	VTN	Method	REQ	VTN	Method	REQ
1	T	13	14	E	2	27	D	3	40	E	1
2	A	1	15	T	2	28	T	2	41	D	1
3	E	1	16	D	5	29	A	1	42	D	2
4	E	2	17	A	1	30	A	12	43	A	4
5	A	3	18	T	10	31	D	1	44	T	4
6	A	4	19	A	2	32	E	1	45	T	2
7	T	2	20	E	2	33	T	1	46	A	1
8	T	2	21	E	2	34	D	3	47	D	1
9	T	3	22	D	1	35	E	1	48	E	1
10	A	5	23	A	3	36	E	3	49	A	6
11	D	2	24	T	3	37	A	8	50	D	2
12	E	2	25	T	2	38	A	2	51	T	2
13	A	4	26	A	2	39	T	1			

Table 3 Requirements Summary by Methods and Tasks

Method	Tasks	REQ
Analysis (A)	16	59
Test (T)	14	49
Examination (E)	11	18
Demonstration (D)	10	21
Special (S)	0	0
	51	147

specifications, for a total of 24,650 requirements on the program that will have to be verified. In some fashion, these 24,650 requirements will have to be assigned to some number of tasks of the five methodological kinds driving the development of the verification plans for system test and evaluation, item qualification, and item acceptance verification actions.

The first verification planning step on a real program is to determine how the requirements in the program specifications should be associated with the verification tasks. The good news is that this must not all be accomplished as a single task, and it need not be accomplished prior to release of a specification. This work is distributed over the span of time required to develop the several layers of entity specifications. We will develop the system specification first, followed by the item performance specifications, parts, materials, and process specifications, and finally the item detail specifications. Commonly, the item qualification verification design will be the first one accomplished in layers, followed by the development of the system test and evaluation design, and finally the detail specifications, delayed until during the design work.

For the one item expressed in Figure 54, there will have to be a total of 51 qualification verification task plans, 51 qualification verification task procedures, and 51 qualification verification task reports of the kinds noted in Table 3. Each of these documents should be formally reviewed and approved, placed under configuration control, and the masters protected from unauthorized change. Each specification will require this kind of verification work response with work related to the system specification flowing into system test and evaluation verification planning, work related to item performance specifications flowing into item qualification verification planning, and the verification work related to the item detail specifications flowing into item acceptance verification planning. On top of all that, there could be thousands of parts, materials, and process specifications on a program.

In general, the work can be focused fairly tightly on the item performance specification being prepared by those immediately responsible for that item, but there must be some system engineers on the program who take a broader view of the verification planning work to view and think across the system, providing an integrating and optimization influence — more on this later.

By now, it is hoped that the reader fully recognizes the large scope of the work that must be accomplished in verifying that a product complies with the requirements that drove the design work. Yes, it will cost time and money to do this work well, but if the product requirements analysis work was done poorly, then it is likely that these errors will be propagated throughout the program, leading to badly stated verification requirements, poorly done verification plans and procedures, design errors, and significant problems in verifying compliance, recognized at a time when the available budget and schedule slack is minimum. It is a reality that systems engineering work done well early in a program will reduce development cost, and most often system life cycle cost as well. So, if you want your enterprise to be a developer of affordable systems that customers long appreciate, do the requirements and synthesis work well; it will lead to affordable verification if the whole is managed well.

It is not necessary that everyone on a program understand how deep and broad the verification process scope can be, but it is essential that whoever has the overall responsibility for the verification work can see it very clearly and how it is partitioned into items, tasks, and responsibilities. This person must be able to interact with those responsible for parts of the overall work set and discuss problems being experienced with clear focus on how they fit into the whole. Depending on the number of items in a system that require qualification, it may be necessary for two or more layers of qualification principal engineers to be appointed. In such cases the principal engineer in each case should be familiar with the verification scope for the items for which they are responsible, from that item downward.

8.3.4.2 The Compliance Matrix

Earlier we discussed how to link product requirements captured in Section 3 of the specification with verification requirements in Section 4 and verification methods, using a verification traceability matrix that includes only the verification strings corresponding to the content of the specification in which the matrix appears. In order to plan the entire program qualification verification process, we need access to every qualification verification string from every item performance specification, so let us build a qualification verification compliance matrix for that purpose by assembling the union of all item performance specification verification traceability matrices. If we are using a database, this will not be a difficult problem. The reader can immediately see how difficult it would be if we were using typewriter technology as in days now long passed. When computer word processor and spreadsheet applications became available, it was necessary to double book the information with the content of the specification in one application, such as a word processor, and the matrix in a spreadsheet application, for example. With a common database, they can both be generated from a single application, removing many sources of errors from the related work and providing for a few other matrices of value in verification management.

Table 4 offers a partial view of a verification compliance matrix that may require some additional fields to support effective management. The table would benefit from inclusion of the paragraph titles, but there was insufficient space here to include them. Clearly, we have added three fields to the verification traceability matrix shown in Figure 28. One could argue that the LEVEL column should not appear in the verification traceability matrix. This would require it to be determined prior to release of the specification, and this may unnecessarily delay release of the specification. The author has no problem with deleting the LEVEL column from the traceability table but including it in the verification compliance matrix. In any case, the LEVEL column permits us to assign a particular verification string of the selected method to a task to be accomplished in association with the item specification from which the string was drawn into the verification compliance matrix, or to a task related to a different specification. The entry corresponding to the first case would be "ITEM" if the verification task will be accomplished at the level corresponding to the specification. The word "PARENT" corresponds to the intent to promote the accomplishment of the corresponding task to the parent item level, and "CHILD" corresponds to the intent to demote the task to a child specification level.

Table 4 Verification Compliance Matrix Example 1

		Section 3			Section 4			
PID	PT	PARA	RRID	PARA	VRID	Method	Level	VTN
A1254	1	3.2.12.2	R346TG5	4.4.23	RpRU73J	TEST	ITEM	1125
A1254	1	3.3.5.1	RfYu37b	4.4.26	R246Tg5	EXAM	ITEM	4033
A1254	1	3.3.5.1	RhLqU7m	4.4.13	R5U765E	ANAL	ITEM	3232

The first change from the verification traceability matrix is the addition of the verification task number (VTN). A task is a collection of related work that can easily be planned, managed, and performed as a single entity – a single test or analysis task, for example. These tasks should be correlated with items in the product entity structure. A suggested VTN identification numbering technique will be offered a little later, but for now we will apply the simple pattern shown in Table 4, where the VTN is unique within the context of the item product ID (PID) with which it is coordinated. In this case, the string A1254-1125 would uniquely identify the task that the first string in Table 4 is associated with.

We have also added the field necessary to identify the item from which the strings came in the product ID (PID) field. A specification part (PT) field tells us whether the string came from the system specification (0), an item or interface performance specification (1), an item or interface detail specification (2), part specification (3), material specification (4), or process specification (5). Remember, the verification traceability matrix appears in a specification for an item so there is no doubt what item or part which the strings included relate to. However, the compliance matrix includes strings from every specification, so we will need columns for item product ID (PID) and specification part. As an alternative, we could also include the PID in the Level column to permit promotion and demotion more than one level to parent or child.

For example, let us say that the guidance accuracy requirement appears in the guidance set performance specification for product entity A1254 and we intend to verify accuracy at the item level using the test method. But also we intend to verify guidance accuracy at the A12 level, corresponding to the whole aircraft in flight containing an avionics subsystem A125, which in turn contains the guidance set. So, in this case the records in the compliance matrix would have the data in the indicated columns of Table 5 with the requirement title omitted for space. We have inserted a new string in the matrix to illustrate this case. Requirement ID R346TG5 is to be verified in A1254 VTN 1125 but also at end item level (A12) in the next string listed in Table 5 If avionics system (A125) testing in a bench configuration were contemplated on the program, we might also have a string for that task linked to a promotion from requirement ID R346TG5. The reader may recognize Table 5 as a verification cross-reference matrix (VCRM) but we would commonly include additional information in that case.

From these two examples we can see some opportunities for duplication that may be desirable or may unnecessarily add to verification cost. The

Table 5 Verification Compliance Matrix Example 2

PID	PART	Sect 3 PARA	FRAG	RRID	PARA	VRID	Method	Sect 4 Level	VTN
A1254	1	3.2.12.2	1	R346TG5	4.4.23	RpRU73J	TEST	A1254	1125
A1254	1	3.2.12.2	1	R346TG5	4.4.23	RpRU73J	TEST	A12	4214
A1254	1	3.3.5.1	1	RfYu37b	4.4.26	R246Tg5	EXAM	A1254	4033
A1254	1	3.3.5.1	1	RhLqU7m	4.4.13	R5U765E	ANAL	A1254	3232

person doing the compliance matrix for item A1254 could end up including three strings for guidance accuracy, one for A1254, one for A125, and one for A12, all motivated by the performance specification requirement ID R346TG5, where in the latter two cases the requirement verification was promoted to higher tiers. At the same time, the person doing the compliance matrix work related to item A125 could build a string for the guidance accuracy requirement at subsystem level at item (A125) as well as demote the verification to A1254. The reader can perhaps see the need to perform integration and optimization work on the evolving compliance matrix, where different engineers are introducing content possibly without any kind of conversation between engineers responsible for the different levels or management guidance. The enterprise could establish some rules to guide engineers working multiple levels, such as "do not promote or demote a requirement that can be adequately verified at the item level." In this case, guidance accuracy might well be verified for item A1254 in a bench test for the guidance set, A125 in a bench test of the avionics subsystem, and for the aircraft A12 in system test and evaluation. There will be cases where it is important to verify a requirement such as guidance accuracy at several levels, and there will be cases where multiple level verification is a case of excessive and unnecessary work.

Table 5 also includes a fragment (FRAG) column to allow us to differentiate between two or more requirements fragments in one paragraph. In order for the fragment identifiers to be effective, of course, a definition would have to reside somewhere identifying the extent of these fragments. While your organization may apply a rule that each paragraph must focus on a single requirement, you will often have to verify requirements that appear in customer or supplier generated specifications, which could include page-long paragraphs containing many requirements. The picture we have painted may appear complex, but it does cover the waterfront. Some simple rules can simplify the verification planning process. We have cited two and the reader can probably think of some others: (1) only one requirement in each paragraph, and (2) do not promote or demote a requirement that can be adequately verified at the item level.

There is a pair of boundary conditions also for the promotion-demotion selection versus accomplishing all of the work at the item level in all cases. In general, it will cost less to verify a complete radio system, for example, than to do so at the item level with separate tasks related to the receiver, transmitter, antenna, and power supply. The system task will require more problems in coordination and management, but if done well, will produce good results at less cost. If all of these items were being purchased from different suppliers, it might be better to verify at the component level, so that the suppliers could be held accountable for proving their product satisfied their contractual requirements prior to shipment.

Tables 4 and 5 include requirement ID columns that draw unique ID strings from a vocabulary that includes all strings that can be formed using a base-60 system, where each character is from the character set formed by all of the ten Arabic numerals, the capital English alphabet letters less the letter O, and the lowercase English alphabet characters less the letter l. The first character is the letter R to indicate it is a requirements ID. While some engineers feel it is adequate to provide traceability between requirements using the paragraph numbers, these often are changed over the course of a program, leading to traceability conflicts. The requirements IDs are intended to remain unique and constant, so if traceability is maintained through them, paragraph numbers may be changed at will with no effect on traceability. One needs a way to uniquely assign these strings and this requires a database system that can automatically take care of that. With the six characters assignable in base 60, the number of unique IDs is 60^6 or 46,656,000,000, which is vastly more than enough on any program.

Clearly, tabular data is useful in verification planning and management. While an organization can get by using a spreadsheet application for these tables, a database application is encouraged. Where an organization is still operating with typewriter thinking in a computer world, it will be very difficult to create and maintain the data needed to effectively manage the process. Traceability and verification links are the big drivers in movement to the use of computer databases for requirements work that began several decades ago.

Ideally, the specification content and verification data would all be retained in a common database that would allow us to assemble whatever matrices we need. In this book we will discuss the use of verification traceability, verification compliance, verification task, and item tables or matrices, and if the program must maintain all of these tables and the content of the specifications independently, errors will inevitably occur.

The verification task numbers (VTN) included could be unique system-wide, or unique within each PID. The VTNs included in Tables 4 and 5 were chosen to fit into the space available. Shortly, an attempt will be made to offer a unique task numbering scheme.

In many organizations, management is offered the verification compliance matrix as a management tool, even though it is not well suited to this purpose. It is the right work product for design of the verification process, but the wrong tool for management of it. Therefore, the verification compliance matrix should be supplemented with a companion table (from a common database, ideally), called a verification task matrix, which defines the verification tasks in terms of plan and procedure reference, responsible engineer, planned and actual plan and procedure release dates, planned task start and complete date, and current status. These matrices should be published periodically in a verification management support report. Figure 55 summarizes the documents related to the verification process, emphasizing qualification.

The verification traceability matrix in each of the specifications should remain fairly stable throughout the development process. If one finds that including the LEVEL column in the verification traceability matrix results in excessive changes to the specifications after they have been first released, then probably the LEVEL column should be moved to the compliance matrix and that action made part of the verification design process. The verification compliance and task matrices will obviously change as the program progresses and status changes, so it may be preferable to prepare them as separate documents and reference them in the matrices of a verification management support report. This is especially useful where the customer is a DoD component or NASA center and contractual requirements include periodic delivery of these data items. This avoids bumping a program verification plan revision letter each time the management matrices change. The author encourages the inclusion of the compliance, task, and item matrices in the verification management support report published periodically, with a copy made available to the customer. This information would provide the customer with references to verification documents and status of the verification work, which would permit the customer to request specific documents for review or, if the documents are sent to the customer as contract data requirements list (CDRL) items, to focus on those of interest as they become available.

8.3.4.3 Task and Item Matrices

The verification compliance matrix is used by those responsible for planning the verification work to map all of the verification strings from all of the specifications to some finite number of verification tasks of the six methods indicated in the matrix (the five plus none). The verification requirements, linked to specific methods, were completed in each case by the specification principal engineer, supported by engineers contributing requirements to the

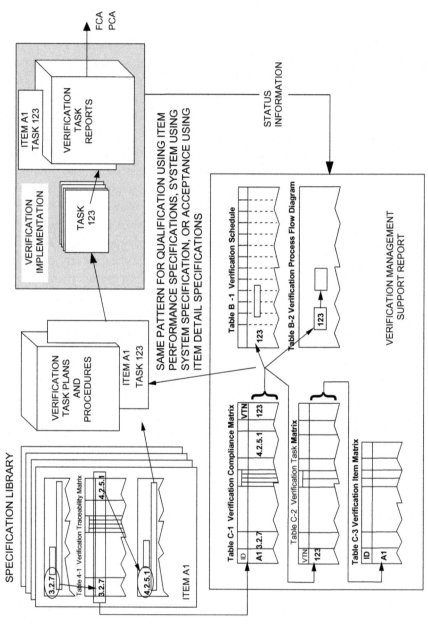

Figure 55 Verification planning and implementation management documentation.

specification. The engineer(s) responsible for overall verification planning must now transform those many strings into methodological collections of strings that can be accomplished efficiently as tasks, and the total collection of tasks for an item completely implements the verification work for all item requirements.

This work is actually a creative design process and, to be successful at it, one must remember the goal is to collect good evidence of the degree of design compliance with the content of the performance specifications affordably. We want the best evidence we can afford. Another point of interest is that the work of the individual specification principal engineer was accomplished focused on that one item. The verification planner may find that it is possible to integrate and optimize on that work at the higher level. If improvements that change the content of the item verification traceability matrix are discovered, these changes should be discussed with the specification principal engineer, and any changes agreed to should be included in the specification. Avoid making any verification process changes that are in disagreement with the content of the specifications that have not been coordinated with the item principal engineer. It is important to maintain good traceability throughout the program time span.

One approach to mapping verification strings to verification tasks would be to consider each string a separate task, as noted earlier. This will result in a very costly and complex undertaking. If the system requirements are documented in 150 specifications, and on average each specification includes 80 requirements, then this would result in 12,000 tasks. If the average cost of a verification task were $500, then the total verification cost would come to $6 million. If through good verification process design we could map those 12,000 requirements to 1200 tasks (an average of 10 requirements per task), and the average cost per task crept up to $1000 because the tasks were more complex, the total cost of the verification program would be reduced to $1,200,000 – that is, about a fifth of the previous cost.

The process of collecting verification strings into tasks is commonly done based on past experience of people who have developed a talent for doing this work. The guiding principle is that one wants to identify a minimum number of tasks, each one of which requires a specific set of supporting resources in terms of materials, equipment, facilities, and personnel. Cost and schedule problems often occur as a result of process transitions, so by minimizing the number of significant transitions we encourage lower cost and schedule risk. There is an opposite extreme as well, of course, where we try to collect too much activity into a single task, making it unlikely that we will actually be able to bring all of the resources together quickly when it comes time to execute the task. There would be a tendency to acquire the resources as early as possible and then focus on the remaining ones that are more difficult to acquire. In the process we will have maintained an inventory of some of the resources acquired early that are not being productively used. If, for example, we have separate thermal, vacuum, and vibration test chambers in our environmental test lab, it would make sense to identify separate verification tasks coordinated with each lab instrument rather than requiring some form of consolidated use of these separate chambers. If, on the other hand, we possessed a combined instrument, it would make sense to combine the verification tasks for the related requirements.

The simple verification tasks involve a single application of a planned process no matter the method to be applied. Where there is an expected variance or tolerance in a measurement, we may have to apply statistical theory to determine how many repetitions of an observation are necessary to assure an acceptable confidence level as well as the measurement value range of acceptability. In some cases, it may even be necessary to test multiple articles to acquire a statistically valid data set.

The verification planner studies the content of the verification compliance matrix for a given item from the perspective of the four methods and sorts the verification strings of each method into collections that will become verification tasks. If the data is in a database, then the planner can set the filter for the item of interest and also for a particular method if desired, to allow him or her to focus on strings of interest. The planner can sort the data on method with the filter set to the item and make task identification within each method collection based on a perceived common set-up and supporting resources needed. The outcome is a set of verification tasks with all of the verification strings for a given item mapped to it. When all of the item planning is complete and the consequences of any promotion and demotion handled, the verification tasks are all identified for the qualification process. It might be useful to include several analysis fields in the compliance matrix, just to support study of different ways of organizing strings into tasks.

We then have to populate a verification task matrix with the results of this work, an example of which is shown in Table 6. Task numbers are added

Table 6 Verification Task Matrix Example

PID	VTN	Method	Principal	Plan	Procedure	Report	Status
A1254	1125	Test	B. Jones	58-41230	58-42231	58-43232	

to this table as they are identified and are linked to the content of the fields noted. A status field would also be useful, and the reader can think of several others. As the document numbers are identified for plan, procedure, and report for each task, they are added to the matrix. Planned date and budget columns could be included and it would also be helpful to be able to capture actual data as well. The resulting matrix provides managers what they need to manage the evolving process. It names the people responsible and allocates to them the available schedule and budget resources they require to accomplish the task. All of the fields noted are not shown in Table 6 in the interest of space on the page. The performance of these principal engineers can now be managed relative to these resources.

The qualification planning work up to this point has been focused on the top-down allocation of verification strings, each composed of item product requirements, verification requirements, and methods of verification, to tasks. Now the focus must shift to the verification tasks formed of work required to satisfy a particular collection of strings oriented toward the items to be qualified and methods of verification to be applied. This remaining work is part of the bottom-up planning process involving preparing verification task plans and procedures for each verification task and moving toward a condition of readiness to begin the planned work by acquiring the resources identified in the task plan.

The verification compliance matrix identifies all of the requirements that must be covered in the task plan and procedure for a given VTN. The task matrix identifies the engineer responsible for crafting the task plan and procedure and provides other needed management data. The responsible engineer must prepare a plan and procedure that will produce compliance evidence for all of the verification strings mapped to the task. Where the method is test, a plan will be developed by the principal engineer that calls for a series of task steps involving test stimuli, responses, and decision criteria; a set of test resources (test equipment and apparatus some of which may have to be created especially for this program) is identified; and cost and schedule constraints are identified and balanced with available resources. Demonstration is similar except special test equipment is generally not called for. Examination and analysis will involve human observation of the product and its representations coupled with human thought processes to reach conclusions about the comparison of product design and the driving requirements.

The compliance matrix is useful to the verification engineer planning the verification process by mapping verification strings to VTNs, thus identifying the tasks that the verification process shall consist of. But, the compliance matrix is the wrong view for management, because it is full of detailed data and management commonly needs information cut at a higher level. The verification process is going to be managed at the task level, where a task commonly depends on one collection of related resources, the same test lab and personnel, and all under the same organizational responsibility. It is the tasks about which we will prepare verification plans and procedures as well as verification reports. So, it is the task level status that must be reported periodically to management.

Multiple rows in the compliance matrix (all of those mapped to a particular task) become one row in the task matrix. Rather than try to map all of the plans, procedures, and reports to the requirements in the compliance matrix, we should use the task matrix for that purpose. It will obviously entail less data entry work and reduce data maintenance cost. When using a well-designed database for verification data management, these two matrices simply become different views of the data. All of the qualification verification work associated with a particular item in a large system undergoing verification will take up several rows in the task matrix, one for each task for that item.

A fourth matrix can be of value on a large program involving many items that require verification work. An item matrix provides for the highest level of generality in management reporting, where one row in this matrix corresponds to all of the task rows in the task matrix mapped to that item. Table 7 illustrates an example of this matrix. Where the item matrix is helpful to management is when one is trying to judge whether or not the program is approaching a condition of readiness for a planned item audit. In early September of 2012, according to Table 7, 60 days remain before a scheduled end of item A1254 qualification verification. The responsible manager can check the completion status in the item matrix, which shows 55.3% (68/123) for item A1254. Then the manager can check with the principal engineer (Jones) for the item qualification process and inquire about the remaining 44.7% [(123 − 68)/123] of the tasks. The manager may find that two difficult tests remain incomplete as well as 53 other tasks, causing the principal to make an adjustment in the planned end of item qualification

Table 7 Verification Item Matrix

	Numer of tasks			Date complete		Principal	
PID	Part	Total	Complete	Scheduled	Adjusted	Engineer	Status
A1254	1	1235	68	11-01-2012	12-01-2012	P. Jones	In Work

by one month. As a result, the manager might make an appeal to the customer to delay the FCA by one month. Alternatively, the manager could allocate more budget to the task, encouraging earlier completion. The reader may be able to think of other applications for the item matrix. It would be helpful to include planned FCA dates and readiness state in the item matrix or have them be accessible in the database within which the matrices are maintained.

Figure 56 illustrates the connectedness of all of these matrices. The figure supports the integrated documentation approach suggested in previous editions of this book, which will be discussed later as an option. The compliance, task, and item matrices, in addition to the verification schedule and a

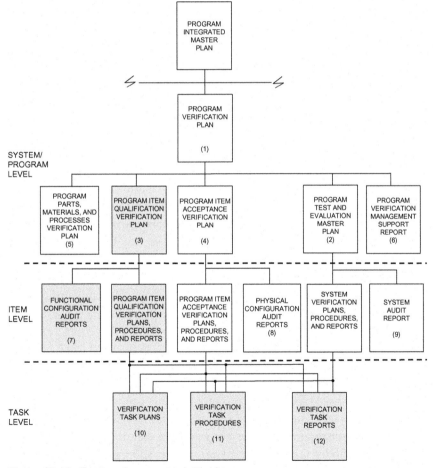

Figure 56 Verification document relationships.

flow chart or network diagram, can be collected into an integrated verification management support report. The contents of this document may never have to be actually published as a paper document except for particular milestones; rather, interested parties should be able to access the parts of it they are interested in on a computer workstation. The document will change frequently to keep track of evolving status, making a paper-publishing requirement quite a chore, especially since it will commonly be voluminous.

8.3.5 Task Principal Engineer Assignment

One of the most critical steps in the planning process is the assignment of the task principal engineer. This person will be made responsible for preparing the task plan, preparing the task procedure, making sure the resources needed to accomplish the task identified in the plan are brought together in accordance with the schedule, managing or performing the task in accordance with task documentation, and reporting upon the task results. The assignment could be the design engineer most responsible for the item design or a specialist familiar with the kinds of requirements being dealt with in the task. If cross-functional teams were employed during the design task, it would be ideal for people who participated on the item team to be selected, provided the teams are funded through completion of qualification.

In any case, whoever is selected as a qualification verification task principal engineer, that person must be given a list of the requirements assigned to that task and be made to understand that the principal engineer must include a table in the task report that links each requirement to the precise location in the report where the evidence of compliance is included, in terms of a paragraph, figure, or table number. It must be easy for anyone reading the report to reach a conclusion about whether the design complies with the requirements and it is seldom that this is the case in practice. If one wants to make verification successful as well as affordable, the content of this paragraph is the surest way to reach that goal.

8.4 ITEM QUALIFICATION VERIFICATION DOCUMENTATION

The enterprise should have available for deployment to each new program a set of document templates and data item descriptions (DID) for each kind of document that the program will have to create. In the case of the verification activity, this includes a task plan, a task procedure, and a task report series for system test and evaluation, item and interface qualification, and item and interface acceptance. Parts, materials, and processes will commonly involve a mix of supplier formats, though the enterprise should possess a template and DID format for each of the three kinds of documents for parts, materials, and processes it develops for use in its own products and sale to other enterprises.

8.4.1 Qualification Documentation Summary

Program specifications provide the input to the verification planning process in four formats: (1) the system specification driving system test and evaluation verification, (2) item and interface performance specifications driving item and interface qualification verification, (3) item and interface detail specifications driving item and interface acceptance verification, and (4) parts, materials, and process (PMP) specifications, driving PMP procurement and verification.

We must have a way to publish all of the program verification planning, and the author encourages the use of a document he calls a verification management support report, containing the verification compliance matrix, verification task matrix, verification item matrix, verification schedule, verification flow diagram (by whatever name), and some text explaining the overall process. The details and mass of this planning data will appear in the verification plans and procedures prepared for each task and listed in the verification task matrix. Past editions of this book have encouraged the use of integrated verification plans, procedures, and reports documents, but after many years in aerospace the author has never observed this approach employed and has concluded that it does impose an unnecessary added burden on the program that will make it more difficult for everyone contributing to verification success to provide their documentation inputs.

Therefore, the author's preferred approach has shifted back to a more traditional one involving a separate plan, procedure, and report for each task. In addition, a program should publish an overall verification plan, a development test and evaluation plan, and reports dealing with any functional configuration audits (FCA) or physical configuration audits (PCA) accomplished. The author wishes that this list comprised the whole of the verification documentation story, but is sad to report that on many programs there is a tremendous amount of documentation between customer and contractor dealing with the ongoing status of the verification process. This often happens because of a lazy customer, and sometimes because of customer distrust of contractor verification performance, earned or otherwise from past performance. The documents listed here should be adequate for any verification process, no matter how complex the system, but the customer must be staffed and be willing to actually read the verification documents and the contractor must publish them such that they may be easily understood – especially the reports. In the next few chapters, we will discuss the plans, procedures, and reports with special emphasis on encouraging understandability.

8.4.2 Verification Documentation Suite

Verification documentation done well can pose a significant cost burden to a development program. If it is done badly, the cost of preparation may be less but the overall program cost will generally be terrible in its effect and value. The author was, in fact, motivated to write this third version of this book because of the many overruns he had observed on program verification work and his failure to deal with it well in the earlier editions. On a large program, there will be many documents of several kinds and it is a challenge to prepare them all well and affordably. Unfortunately, too often this work is not done well on programs and the result is a real burden on the program. The development work to prepare these documents should have been pre-ceded, of course, by the well-done preparation of another series of docu-ments, the program specifications, and often verification problems encountered are driven by prior poor work in preparing the specifications. Before focusing on item qualification verification task plans in this chapter, let us summarize the full verification documentation burden illustrated in Figure 57. Documents shaded yellow are applicable to item qualification. Those documents can be grouped for convenience in discussing them at the program level, item level, and the task level. Program level verification documentation using the author's titles includes those depicted in Figure 57 and listed with brief explanations following.

(1) Program Verification Plan – A program level verification plan immediately subordinate to the program plan, covering the full sweep of verification efforts including system test and evaluation, item quali-fication, item acceptance, and PMP verification. This document should not include the details but rather the overall pattern in which the sub-ordinate documents assemble and provide the details. A top-level flow diagram and schedule should be included, telling in what order and when the work is intended to be accomplished. Budgets should be assigned for the four major components of the verification work: devel-opment test and evaluation (DT&E), item qualification, item accep-tance, and PMP. The author's preference is for a program to prepare this program summary plan and four subordinate plans for each class,

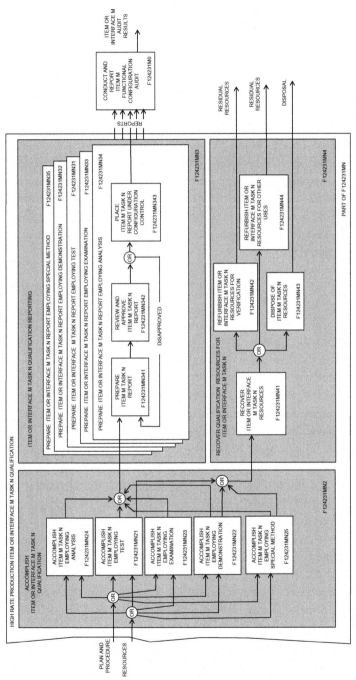

Figure 57 System verification documentation.

but on a small program all of this content could be included in a single program level plan. The content of this document could be included in a multivolume integrated master plan (IMP) and integrated master schedule (IMS), but the author prefers separately published documents.

(2) Program Test and Evaluation Master Plan (TEMP) – This is the master plan for DT&E subordinate to the program verification plan. The DT&E will be accomplished in some number of tasks, each one in accordance with a task plan and procedure within the constraints made clear in the TEMP. While the many tasks that comprise the TEMP will be reported upon in verification task reports, it may be desirable or necessary for a program to publish a closing summary report of the results of the DT&E. Responsibility for each DT&E task should be clearly stated, as it should also be for how the work is to be managed, the name of the principal engineer, task budget assigned, schedule constraints, and facilities required.

(3) Program Item Qualification Verification Master Plan – This document provides an expansion of the program verification plan content as it relates to the item qualification component.

(4) Program Item Acceptance Verification Master Plan – This document provides an expansion of the program verification plan content as it relates to the item acceptance component.

(5) Program Parts, Materials, and Processes Verification Master Plan – This plan covers the union of a PMP verification plan, preferred PMP policy, and planning and results of efforts to manage parts and materials availability over the system life cycle.

(6) Program Verification Management Support Report – This report provides the overall verification schedule for the three verification phases and should be updated as program circumstances unfold. The report should include copies of the current verification compliance, verification task, and verification item matrices. The content of this report will change frequently and should be updated at a pace determined by program management. This document is the primary way that those who are doing the verification work can communicate the current status to those responsible for managing the work. Ideally, the component documents found in the report could be accessed via a computer for status of the moment between formal updates. This document could be released as part of the program management plan or in the form of appendices to that plan. The author prefers to release it as a separate document referred to in the program verification plan,

because it will pass through a different frequency of change than the overall plan.

(7) Verification work is accomplished oriented toward the requirements contained in a system, item, or interface specification. When the verification work is completed for item qualification or first article acceptance, an audit is often held to reach a formal conclusion between customer and contractor on the degree to which the item complies with the requirements contained in the related specification. Item level documentation includes:

(a) Item Qualification Functional Configuration Audit Report – Report identifies all of the item qualification tasks and gives document numbers for plans, procedures, and reports in each case. The report identifies any requirements where initial compliance was not established and gives a chronology of the actions taken to reach compliance.

(b) Item Acceptance Physical Configuration Audit Report – Audit occurs after completion of first article manufacture is complete. Report lists all acceptance tasks coordinating task plan, procedure, and report numbers. Provides results and offers a conclusion about degree of compliance.

(c) System Audit Report – The author encourages that a formal audit be held at the end of development test and evaluation to reach a formal decision on the degree to which the contractor has been successful in showing that the system design complies with the content of the system specification, and identify what, if any, actions the contractor must take to attain final customer approval of the development test and evaluation program results. Ideally, the customer would hold a similar audit at the end of operational test and evaluation (OT&E), allowing contractor attendance.

(8) A verification process is designed by identifying a collection of tasks for each item in the system at some level of indenture, including the system as a whole, such that the accomplishment of the tasks identified for a given item will collectively provide clear evidence of the relationship between the content of the item specification and the characteristics of the design and/or manufacture of the product entity. These tasks will all coordinate with one of five methods: (1) test, (2) analysis, (3) demonstration, (4) examination, and (5) special. One other method that should be recognized on a program is "none" where a specification paragraph has no significant content, but in these cases there is no need for

documentation of any kind. In the case of each task, three kinds of documents must be prepared: (1) a plan telling what must be done, (2) a procedure telling how it must be done, and (3) a report telling what happened and how the results should be understood. While this book encourages separate plans and procedures, a combined document often makes a lot of sense. These documents are further influenced by which of the four verification classes the task relates to: (1) item or interface qualification based on the content of the performance specification, (2) system test and evaluation based on the content of the system specification, (3) item or interface acceptance based on the content of the detail specification, or (4) parts, materials, and processes verification based on the content of the parts, materials, and process specifications. So the template for one of these documents will fit into one of 60 formats as a function of the three kinds of documents (plan, procedure, and report), five methods (test, analysis, demonstration, examination, or special), and four classes (system test and evaluation, item qualification, item acceptance, and PMP). If we accept PMP as three different formats, then the number jumps to 90.

(9) Previous editions of this book encouraged the development of three integrated verification documents: (1) an integrated program verification plan, (2) an integrated program verification procedure, and (3) an integration program verification report. Over the period of 20 years these books have been available to the public, the author has seen no evidence of anyone following this encouragement and he has concluded that the integrated document pattern is a case of overreach and is too hard to accomplish. Therefore, this edition encourages the development of these documents independently for each verification task and treats the integrated document approach as an option. Thus, it is intended that for each verification task identified in verification process design, a program would prepare three types of verification task oriented documents:

(a) Item Verification Task Plan – A task plan will be required for every task identified in the system test and evaluation, item qualification, and item acceptance plans. Many of these will be very simple and others very involved, but in all cases should identify: the resources needed, budget and schedule provided, the name of the principal engineer, a list or table identifying specifically what requirements in the related specification the task must determine the degree of design (system and qualification) or manufacturing (acceptance) compliance with. Appendix A1 offers a template for a qualification task plan and Appendix A4 offers an item acceptance task plan template. These templates imbed the procedure into the plan document.

(b) Item Verification Task Procedure – A task procedure will be required for every task identified in the system test and evaluation, item qualification, and item acceptance plans. Each procedure tells how to accomplish the work required by the corresponding plan. As a consequence of performing the procedure, specific results are observed and recorded telling to what extent the product complies with the requirements assigned to that task.

(c) Item Verification Task Report – A task report describes the results observed when the plan and procedure have been implemented in terms of the degree of compliance established. Appendix A2 offers a qualification task report template and Appendix A5 offers an acceptance task report template.

8.4.3 Item Qualification Verification Planning Documentation

An enterprise should have available for use on programs a set of generic templates for all of the documents identified in Paragraph 8.4.2 and also a data item description (DID) in each case. The template simply provides a document structure in terms of its organization, using an outline. A DID provides information about how to translate a template into a specific program task document. In the most extreme case, an enterprise would need (3 verification classes) × (3 document types) × (4 verification methods) = 36 verification task templates and 36 DIDs, one-third of these for item qualification. It is true that at the task level the item qualification and system test and evaluation documents will have a very similar structure and it is possible that one could get by with 24 of each kind of document, but there are some unique facets to system test and evaluation, so the author is not willing to surrender on this point too easily. Alternatively, a program could combine item plans and procedures into a single document in each case resulting in two document types or 24 kinds of documents of each of the two kinds. In this case, if we also combined item qualification and system test and evaluation under one type, it would be possible to employ only 16 kinds of documents.

In the worst case, we would have to maintain a large number of templates and DIDs, but the author believes it is possible to use the same basic document structure regardless of method employed. The result calls for three kinds of documents (plan, procedure, and report) for three classes of verification (system, item and interface qualification, and item and interface acceptance). System acceptance tasks are not recognized in that the whole system is not really manufactured in one task. Also item qualification task documentation is effective for system test and evaluation tasks. In this chapter we deal with qualification, so will only have to deal with five templates and DIDs for the three kinds of documents. Chapter 9 will focus on system test and evaluation and Chapter 10 on item acceptance verification documentation. It is entirely possible we will find that we have to provide different lower tier content as a function of the method in one or more of the five kinds of documents in each verification class, but it is possible to frame the content in a common structure. Appendix A provides a set of data item descriptions for verification documentation. The reader may transform this set of DIDs into a set of templates by deleting the text from the provided DID, leaving only the paragraph numbers and titles.

As noted previously, this book makes an attempt to offer a generic plan template that can be implemented for a task using any of the five methods. To gain confidence that this can be done, the author has included a flow diagram for each of the five methods. Most of the unique aspects of planning relative to method employed will come in plan Paragraph 4, dealing with accomplishing the task. As noted earlier, an enterprise could choose to depend on the list of templates and DIDs listed in Figure 33 or 34. The author believes that the list in Figure 33 would be preferred, but it is a difficult standard to complete. Once an enterprise masters this level of detail, they can begin to incrementally expand the document inventory to comply with the Figure 34 listing. This book offers a set of task document templates coordinated with Figure 34 in Appendix A. Appendix A1 provides a combined item or interface task plan and procedure template by including the procedural steps in the plan.

A plan is a list of steps to be performed to accomplish a task or achieve an objective, captured in a text document, data set, or graphical image that also identifies the timing of steps if important to the objective and a clear description of the resources needed to accomplish the task. The objective in each case is to prepare a document that will produce true evidence of the relationship between the requirements contained in the item performance specification allocated to this task and the design features of the item. If the item performs or is characterized so as to comply with all of the content of the specification allocated to the task, it shall be said to be qualified relative to, or comply with, those requirements.

Our plan begins with administrative information about the task. Paragraph 2 deals with needed personnel and in many cases this will involve only the task principal who will actually do the work as well as plan it. If other personnel are needed it is important to make a solid case for the cost because management review of the plan will require a clear rationale for the cost before approving the plan. Paragraph 3 provides for the identification of other resources needed for the plan including tools, equipment, materials, and computer software applications. Qualification will often involve manufacture of product articles to support testing and these resources are of great value to the program. The plan should tell how the program will gain maximum benefit from these resources during and subsequent to their use in qualification.

Paragraph 4 offers the procedure for accomplishing the task and it is here in the task procedural data that lower tier guidance may require some methodological differentiation. Notes, cautions, and warnings should be included that will emphasize a need for alertness on the part of the person implementing the procedure to specific concerns. Finally, Paragraph 6 covers closing out the task.

8.4.4 Accomplish Item or Interface M Task N

Figure 58 illustrates the portion of the verification task dealing with accomplishing the task, reporting, auditing, and recovery of task resource materials. This activity tags onto the end of the activity shown in Figure 47. As in the case of developing task plans and procedures, there will be many tasks being accomplished at any one time. A task matrix listing all of the tasks, telling who the principal engineer is in each case, and the budget and schedule assigned is needed to manage this activity well. Table 5 offers one line from a verification task matrix that provides managers with this information, recognizing that on a real program of any size this will consist of thousands of records.

The principal engineer and/or personnel assigned to accomplish the task with an approved plan and procedure will have to follow up on planned acquisition of task resources listed in the plan. This may include a product entity that will be subjected to test or demonstration or it may only require access to engineering design information. It may also require gaining access to the personnel needed to accomplish the task, where there may not be sufficient personnel assigned to the program at the time. Verification occurs late in a program and, if the program has not been managed well, it may be in an overrun condition, making it difficult to gain management approval for

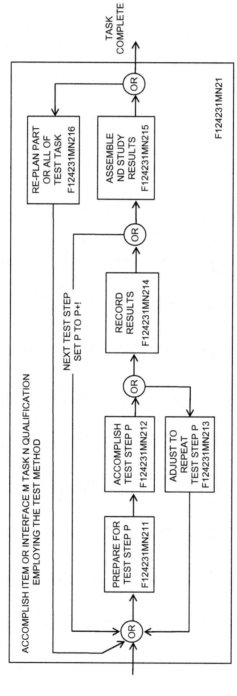

Figure 58 Qualification item or Interface M Task N accomplishment and closing activities.

spending money, even though the task may be in the program plan, budget, and schedule. It is bad form to simply throw up one's hands and blame failure on early program management failures, however valid the claim might be. As painful as these situations may be, it should be recognized that they are also great opportunities against which to test one's engineering and leadership skills with a high degree of management attention to one's skills or lack thereof.

Accomplishing a task should be as simple as a qualified person following the approved plan and procedure. Life is often not simple, of course. The person assigned may not be qualified to accomplish the work. If true, someone in the management path must identify this problem and seek to correct it as early as possible. Some possible solutions include: (1) change the assignment so as to provide someone with the skill and knowledge needed to do the work, or (2) use the opportunity to upgrade the engineer's skill and knowledge in a kind of train-while-doing fashion, so that they will in the future be able to do this kind of task. The latter may call for a little more cost as well as a small schedule hit, so management should be included in the decision-making process.

After completion of a qualification task that subjected a product entity to test or demonstration actions that applied life-draining stresses, the article cannot be shipped to the customer. There are applications of value for these articles, however. The first possibility is that some of the qualification work will have to be repeated subsequent to a design change or recognition that the plan and procedure must be changed after the fact. In these cases, the article may have to be refurbished and pass an acceptance test to ensure that it represents the condition of a delivered article. Alternatively, the article may have to be replaced by a newly fabricated unit, which may include design changes possibly calling for some changes in the plan and procedure.

In any case, a post-qualification article may still possess useful capabilities. Even a black box that is inoperative can be used in item remove and replace verification tasks, where one times the removal and replacement of the item and compares the time with that required in a maintainability requirement. The item can be subjected to mass properties inspections involving weight, center of gravity, and dimensional measurements. A jet engine that has been exhaustively tested to the extent that it no longer can be operated retains value as a training device and if all else fails can be retained by the enterprise program office on a suitable stand as a static display or presented to the customer program office. Continue to look for residual value for all of these items until they retain no more program value or have been disposed of.

Each qualification verification task will employ one of five methods: test, demonstration, analysis, examination, or special. An attempt has been made in the plan template discussed in this chapter to offer documents that will generically fit any of these methods, but there are unique aspects to the application of these methods that will influence our development of task plans, procedures (if separately employed), and reports.

8.4.4.1 Accomplish Item or Interface M Test Task N

Figure 59 illustrates the generic test task process. The principal steps, often taking place in a cyclical pattern, are: prepare for the test, accomplish the test, and record and study the results. In some cases it may be necessary to refurbish resources and dispose of residual material in accordance with special instructions in the plan because they are classified, hazardous, or they retain program value for use in other tasks. If the results do not support the conclusion that the requirements were satisfied in the design, then it may be necessary to make adjustments and retest. A test is commonly accomplished in one or more steps, where all steps require essentially the same supporting resources and personnel. Different steps may also relate to different requirements allocated to the task. The design of an item that will require qualification testing should integrate the qualification test access to signals with that for in-service testing.

8.4.4.2 Accomplish Item or Interface M Demonstration Task N

Figure 60 provides a generic demonstration process flow chart. It is similar to the test task described above. A demonstration, like a test, can take place as a single step or consist of a series of steps, either repeating the same step in a probabilistic data collection sequence or moving on to a different purpose. It is possible to damage or overstress the product in a demonstration, so provisions may have to be provided for refurbishment to support continuing qualification work. Disposal of residual materials may also be necessary subsequent to or during a demonstration.

8.4.4.3 Accomplish Item or Interface M Analysis Task N

Figure 61 illustrates a generic analysis task. The same basic string of tasks is included here as in test: prepare, do, and record. If the evidence does not support the compliant conclusion, we may have to adjust our analytical approach or change the design and subject the new design to analysis.

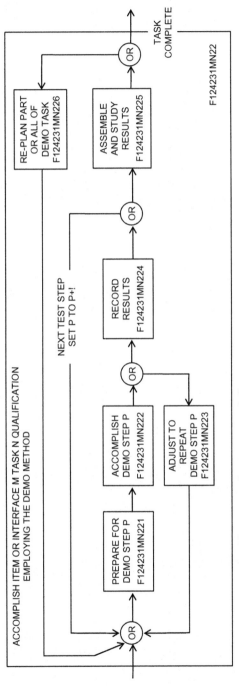

Figure 59 Generic item qualification verification test task flow.

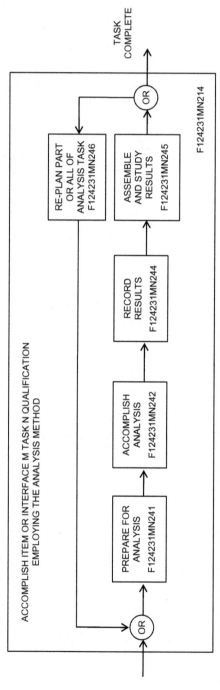

Figure 60 Generic item qualification verification demonstration task flow.

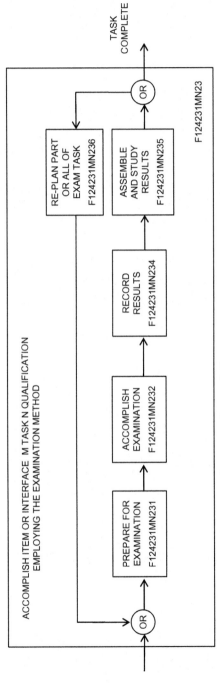

Figure 61 Generic item qualification verification analysis task flow.

8.4.4.4 Accomplish Item or Interface M Examination Task N Examination Task

Figure 62 offers a generic examination process. It is similar to the analysis process in that there is no special-purpose instrumentation involved. This is primarily a matter of one or more humans making observations essentially with their unaided senses.

8.4.4.5 Accomplish Item M Special Task N

The special qualification task generally involves modeling and simulation but can involve any process other than the four primary methods. Therefore, no flow diagram is offered, in that the variations are too numerous to diagram. As an alternative to recognizing the special method, an enterprise could include modeling and simulation under one of the other four methods, such as analysis.

8.4.5 Item Qualification Verification Task Reporting

More programs fail to accomplish an affordable and effective program verification activity because they fail to employ one simple technique that must be implemented in two steps: (1) assign and brief the principal engineers, and (2) report the results of tasks. Earlier in discussing assigning and briefing

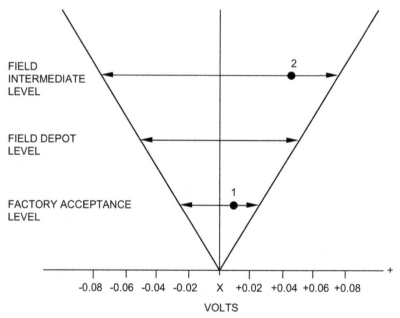

Figure 62 Generic item qualification verification examination task flow.

principal engineers, we said that the principal engineer must be given a list of the requirements for which his or her task and report must provide clear evidence of the degree of compliance demonstrated by the design of the entity subjected to verification in the task. The message must be clearly communicated to the principal for each verification task what those requirements are, and that their final report for the task must include a table listing those requirements, and in each case include a clear reference to exactly where in the report the evidence of compliance is included and whether the design complies or not. These references could be a reference to a paragraph number, figure number, or table reference. Further, the evidence included in the report must be truthful and clearly written, so that any reader would derive the same conclusion.

These are not common characteristics of verification activities the author has observed on many programs. This is a matter that is very simple, but it apparently is not easy to achieve. It requires good supervision of the persons doing the work and good management of the process. It especially requires an effective review and approval of the reports. Those reviewing the report in each case must clearly understand what requirements were involved as defined in the program verification compliance matrix, be able to easily find the evidence of compliance, and be able to understand the degree of compliance achieved by the design from the content of the report.

Each report should pass through the three activities shown in Figure 58: (1) prepare the report, (2) review and approve the report, and (3) place the approved report master under configuration control such that it cannot be changed without a review and approval of any changes. Function F124231MN3 shown on Figure 58 involves the preparation of a report for each qualification task for item or interface M employing one of the five methods. Appendix A4 provides a template for an item or interface M task N qualification report.

In some cases, these reports may be as simple as a paragraph of text, while in other cases may involve a multivolume document. These reports should be carefully reviewed internally for inclusion of the traceability matrix noted earlier, clarity of reference to contained evidence relative to specific requirements, and the correctness of the content in terms of the truth and believability of the evidence cited. As task reports are approved, report masters should be placed under configuration management control, meaning that they cannot be changed unless the changes are subjected to the same kind of review that the complete document previously experienced.

The contract may require that these reports be supplied to the customer and even approved by the customer. The contract under which the development work is accomplished may require the contractor to stage a functional configuration audit (FCA), activity F124231M6 for each item M subjected to qualification during which the program provides to assigned customer representatives copies of the task reports that are studied, possibly stimulating questions directed to appropriate program personnel. The FCA can be accomplished on each item independently, run as a series of FCAs, each grouping a set of items together, or as a single FCA covering all of the items. As a result of this audit, the customer will reach a conclusion on the degree to which each item complies with the content of the item performance specification.

Often these FCAs will be run in terms of some number of requirements from the item performance specifications selected by the customer, for which the developer must produce evidence of compliance as reported in qualification task reports. It is possible that the customer may require a formal presentation of the qualification evidence included in reports they have reviewed.

The customer may challenge developer evidence of compliance. In some cases, the developer will be able to respond so as to change the customer's challenge. In other cases, the customer may be correct, leading to some form of corrective action. Possible actions could include: (1) change the specification such that the evidence shows compliance, (2) a design change followed by partial requalification, or (3) customer acceptance of a deviation for some period of time after which a change will be implemented that results in at least temporary compliance. The specification change option is not as bad as it sounds. If the requirement in question was item failure rate, it may be possible to reallocate parent item reliability (failure rate) such that another item absorbs a lower value that it can meet and the failed item is assigned a higher value than it demonstrated. In the latter case, technical data may have to be changed to reflect a lower performance or payload capability when using the unit with the deviation and subsequently be changed to reflect full capability with a scope depending on whether all earlier end items were updated or not. When an FCA is completed, the developer should be careful to request a customer program office formal notification of completion of the qualification action.

8.4.6 Item M Qualification Integration and Optimization

8.4.6.1 Assuring Item M Qualification Completeness

The complete qualification story for any item will be captured in some number of reports covering the work accomplished, in some combination of tasks of the five methods. Each of these tasks will have been assigned some number of requirements from the item specification such that the complete content of the specification is included in the aggregate. It is important that the results of each task be transferred to the compliance matrix such that, when all item qualification is complete, the matrix will clearly indicate the degree of completeness of the work accomplished. If at that time it is obvious that some requirements have not been fully dealt with, then some follow-up work will be in order, either to adjust one or more tasks or to introduce one or more additional tasks, followed by some additional work and reporting.

One area where portions of a requirement can easily be missed is where a requirement has been partitioned into two or more tasks. Perhaps one portion of a requirement will be verified through test, while another part is qualified through analysis. The report in each case should properly claim compliance with that portion of the requirement assigned to the task, but should not imply that the complete requirement has been qualified. Another area where errors can enter into the item M story is where a paragraph contains many requirements. An enterprise should not, of course, create any specifications with this error, but may have to deal with requirements with this problem that were prepared by the customer, a supplier, or an associate contractor. In these cases, the offending paragraph should be partitioned into fragments, where each fragment is treated as a separate requirement.

8.4.6.2 Interface Qualification Verification Integration

In general, all performance, specialty engineering, and environmental requirements are single ended in that they relate specifically to one product entity. Interface requirements, on the other hand, invariably relate to a pair of product entities. A program has two approaches that could be employed in verifying the requirements for these interfaces.

If a program uses interface control documents (ICDs) to define the interfaces, then verification of the interfaces can be best done in association with the ICDs. In these cases, the product entity specifications should not contain any interface requirements, rather the interface requirements paragraph should refer the reader to the ICD listed as an applicable document in Section 2 and referenced in the interface requirements paragraph. Where the interface requirements are included in terminal specification pairs, it may be necessary to verify them in terminal specification task pairs. Often a program does include one or two ICDs but all other interfaces are defined in specification pairs, so commonly both policies will be required. One option in this case is to promote the interface requirements pairs to the parent specification if they are both subordinate to a common parent. If the pair of product entities are not both subordinate to the same product entity and there is no ICD, the problem is more difficult and the verification may have to be accomplished in two separate actions, each coordinated with one terminal specification.

There is a lot to be said for verifying interface requirements captured in a specification pair as two separate verification actions and verifying interface requirements captured in an ICD as a single verification action. The supporting rationale is that, in the first case, there would be two verification strings, one from each terminal specification, and in the latter case there would be a single verification string.

In no case should an interface be defined in both the terminal specifications and an ICD, because it will be very difficult to maintain coordination between the content of the documents over time as changes are made. Whoever reviews an ICD should double check the terminal pair of specifications to ensure that the ICD is referenced and that interface requirements are not included in the pair.

All verification methods can be applied to interface verification. Test is certainly effective because the source stimulus can be detected across the interface in terms of its effects. Inspection or examination can be applied to verify the physical integrity of two entities bolted together. Demonstration can be applied where special test equipment is not required to sense a successful coupling of an interface effect across the plane.

8.5 RECOVER ITEM QUALIFICATION TASK RESOURCES

Subsequent to completion of a qualification verification task, we must first determine if any of the resources used in completing the task have been degraded such that they could not be used in future verification work. If a resource has been damaged and was identified as an expendable resource, then it should be disposed of in accordance with any available instructions. Generally no damage will occur in a qualification verification task employing analysis, examination, demonstration, or special. If a resource has been damaged and has a continuing application in verification work, then a study should be undertaken to determine the kind and extent of damage, followed by a determination of cost and schedule of accomplishing the restoration, and a decision reached on whether or not the work should be undertaken or the residual resource should be disposed of.

Product articles that have been subjected to qualification have considerable program value if they remain operable, but this can be so even if they are no longer operable. Throughout a program up through qualification work, engineering will have created and maintained many test apparatuses containing equipment developed as the design was maturing, using breadboards and circuitry built in engineering laboratories. The intent in each case is that these devices would function as the final design was intended to, but there may be significant differences, allowing them to be adjusted to some extent for evaluation for different scenarios. Packaging of these test devices may also be significantly different from the final design. Product surviving qualification can be introduced into test devices that have a continuing application, causing the results derived from these devices to be more representative of the final product.

While the author was teaching a series of systems engineering courses at the facility where the USAF/Lockheed Martin C-17 was developed, the engineer responsible for the Iron Bird test stand gave the author a tour of his facility. The Iron Bird was used to develop and test the flight control system and had to faithfully reflect the final design of the system in terms of hydraulic line length and many other features. In that the building was nowhere near the size of the C-17 laterally or in height, the hydraulic lines running back to the empennage were rigged on a rack looping back and forth to build up length and then run up to empennage actuators located on a platform on what would have been the third floor of this one-story building.

The engineer had had to use some actuators from different aircraft initially, gradually molding the Iron Bird into a more representative system and eventually obtaining items from qualification that still had life and in some cases had been refurbished. While sitting in the pilot's seat, the author asked the engineer when he expected to obtain the final controls for the pilots, since this was a transport and was then using a stick that one would expect to find in a fighter aircraft. He replied that it was the final design to use a stick rather than a wheel type control.

As noted earlier, there remain a lot of useful applications for residual qualification items no longer operable. These include supporting engineering-related verification work such as mass properties measurements, logistic-related verification work involving mean time to remove and replace, and customer training involving nonpowered practical work.

If all else fails, marketing may have useful applications for inoperable units that can be mounted on a stand and presented to the customer program office for display in the program office. The contractor must be careful in doing this, however. In one case, the contractor thought that the unit would look so much better if it were gold-plated. Congressmen viewing it did not appreciate that gesture and drew the wrong conclusion about prior frequent contractor program office demands for more budget. This kind of message ends up being delivered back to the source, of course proving that all good deeds receive an appropriate punishment.

8.6 ITEM QUALIFICATION VERIFICATION MANAGEMENT AND AUDIT

Upon completion of all of the qualification verification work and approval of all of the reports on this activity for a given item, customers may insist on reviewing the evidence and reaching their own conclusions on the degree to which the design has satisfied the previously approved requirements. This audit of the results of the qualification process is called a functional configuration audit (FCA) in DoD. A functional configuration audit may be accomplished for each item, or a program may hold individual FCAs or group them in various ways, depending on the needs of the program. Where there is more than one FCA, there should be a system FCA at the end to evaluate the results of all of the individual FCAs and assure that all issues have been addressed.

Each of the item verification processes must be managed by an assigned authority, a team or an individual within the context of an overall verification process management plan. If this process corresponds to a team organizational responsibility, then the team should be the agent, and the team may appoint someone to act as its verification lead engineer. Given that there is more than a single verification task in the item verification process, that lead engineer should coordinate all of those tasks and ensure that the goals of each are satisfied in the work accomplished. This team verification lead engineer may have to accept the responsibility for more than one item task group, leading to a multiple-item responsibility group. Otherwise, the item task and item responsibility groups merge for a single item.

Figure 58 illustrates the internal complexity of a single item qualification verification task that may be composed of many verification tasks, applying the five methods. As noted previously, some of these methods blocks may be voids for a particular item. The figure shows all of these tasks in parallel once again, but on a particular program the order probably will have to follow some series parallel network of tasks driven by task dependencies. Analysis tasks may link to particular test tasks, the output of which is needed in a subordinate analysis task.

The reader will note that the author chose to organize all of these tasks within the order LMN for item, method, task. This has the effect of distributing all test tasks rather than pooling them into one contiguous group. People in a test and evaluation function will not prefer this method of organizing the tasks for this very reason, preferring instead perhaps the order of method, item, task. The author will choose this same structure for the IVP outline for plans and procedures. The rationale for these choices is that we should make every effort to align the management data in accordance with the team structure on a program, to reduce the intensity of the relationships that cut across this structure and strengthen the links that bind together the management strings in accordance with the team structure. Since the author has encouraged product oriented cross-functional teams, it makes sense to follow through with team, or item, high on the verification data organization structure priority.

The result will be that item teams can focus on a contiguous section of the IVP as their responsibility. Other methods of organizing this information will scatter the team responsibility in planning data. If all of this information is captured and retained in relational databases, this will not be an impediment to anyone wishing to focus on only the test planning data, as the database can be structured to apply that filter.

8.7 ITEM QUALIFICATION VERIFICATION VARIATIONS

The preceding discussion actually describes the activities to be expected in one of three possible qualification processes, referred to in this book as high-rate, low-rate, and low-volume, high-dollar sequences. In the high-rate case, described in this chapter, items subjected to qualification verification will be manufactured in an engineering laboratory. These actions are considered to be part of the Systems Synthesis activity (F43) rather than Systems Verification. Subsequently, the program will commonly have to produce one or more of each end item to provide the assets needed for system test and evaluation before the full production capability has been put in place. These will commonly be produced on a low-rate line. Since the items are going to be employed in system test and evaluation, the design should be proven through item qualification so that system testing can be entered with the greatest possible degree of safety for the personnel involved in operation of the test articles.

When the system test and evaluation has been completed, assuming the same contractor continues as the supplier, any changes dictated by system test and evaluation should be adopted and production items built in a high-rate qualification process for delivery. On very large programs such as the Joint Strike Fighter (JFX or F-35), the competing contractors for the demonstration validation phase had to actually build aircraft that were competitively flown in flight test programs. The items in these aircraft had to be qualified prior to entering flight test. The winning contractor then had to begin a production program during which requirements and design changes introduced after the original qualification process had to be reverified and subjected to a new flight test program. Manufacturing for the whole competitive phase for all contractors would have been run at a low rate because only a few aircraft were built by each contractor. The production program would have included a low-rate production process to build the new flight test aircraft, followed by a high-rate production process once the design was proven and a winner selected. This program was further complicated by a need for three different types: (1) Air Force, (2) Navy capable of operating off aircraft carriers with tail hook and catapult interface, and (3) Marine with vertical take off and landing.

A low-volume, high-dollar special case will be covered in a later chapter. Commonly in this case, we cannot afford to dedicate engineering items to the qualification process and have to find a compromise that permits effective qualification while controlling cost. There is also one other case that will be covered later as well – the special case where only a single item, such as a bridge or dam, is being delivered. In these special cases we have to combine qualification and acceptance while not depriving the product of life or reliability.

8.8 ITEM QUALIFICATION CLOSEOUT

When all of the qualification tasks planned for an item have been completed and reports reviewed and approved for those tasks, the program is prepared to host a functional configuration audit (FCA) for the item in question, as noted earlier. The customers may prefer to review and approve each report as it is delivered to them, but in any case what the contractor needs is some kind of enduring statement to the effect that the customer accepts that the item design complies with the content of its performance specification and that production in accordance with the contract may proceed. Appendix A offers a functional configuration audit report template.

Upon completion of all qualification work, the program will possess many product articles that will have completed qualification, and even those that continue to function will offer great value in other roles that the program should take full advantage of.

CHAPTER 9

Item and Interface Acceptance Verification

9.1 ITEM ACCEPTANCE TEST PLANNING ANALYSIS

The qualification work discussed in the prior chapter is intended to take place only once, inspecting each product and interface entity design for compliance with the content of the performance specification for that entity. This activity is scheduled to conclude before the items in question begin production. Successful completion of qualification will assure the customer that the design is compliant, but will not provide evidence of assured delivery of manufactured product that complies. Every article delivered should have been inspected for compliance with the content of the item detail specification, and this is the purpose of acceptance verification. This book is based on programs that prepare both item and interface performance and detail specifications. The system specification, the basis for system test and evaluation, is prepared in only a performance format. It is possible for a program to prepare a single specification for each item in which both performance and detail content is included, which would provide the basis for both qualification and acceptance verification planning. It is also possible that a program could prepare no interface specifications, preferring to include interface requirements in pairs of specifications.

9.1.1 The Notion of Acceptance

The customers of large systems are, or should be, very careful to assure that the products they acquire are precisely what they intended to purchase. Therefore, they should insist on some form of proof that the product does satisfy some criteria of goodness prior to accepting delivery of the product. The requirements that should have driven the design effort are a sound basis for at least part of that criterion. We have seen in the prior discussion of qualification how the performance specification requirements are the foundation for the whole qualification cycle. Those requirements were contained in development, Part I, or performance specifications, depending on what standard you refer to. The specification that contains the requirements we

System Verification
http://dx.doi.org/10.1016/B978-0-12-804221-2.00009-7

249

are interested in for acceptance of each article is variously called a product, Part II, or detail specification. The performance specification content should be solution independent to the maximum extent possible, in that this specification should describe the problem to be solved. The detail specification requirements should be solution dependent in that they are describing features of the actual design that will be used as the basis for determining acceptability for delivery to the customer.

The testing work done in association with development (i.e. qualification) is normally accomplished only once unless there are significant changes that invalidate some portion of the results, such as a major modification, change in the source of an end item or major component in the system, reopening of a production line with possible subtle differences in the resultant product, or an extension of the way the product is going to be used operationally not covered in the initial development. Acceptance testing, on the contrary, is commonly accomplished on each article, at some level of indenture, that becomes available for delivery, and successful completion of the acceptance inspection is a prerequisite of delivery to the customer and customer payment in accordance with a contract.

We must recognize a fundamental difference between item qualification and item acceptance throughout the development effort. All verification work carried out under qualification would not only discover whether or not the design complies with specification content but also how much margin there is in the design. This means that qualification testing will commonly leave little product life left and no item undergoing qualification should be delivered to the customer, with the exceptions covered under one-of-a-kind and low-volume, high-dollar programs. All items subjected to acceptance verification must retain full life.

9.1.2 Where Are the Requirements?

A program may use two-part specifications, one-part specifications, derivative specifications, or some combination of these. In the two-part scenario, the requirements that drive the acceptance inspection process are in the Part II or detail specifications. In a one-part specification, the document includes both performance and detail requirements that drive qualification and acceptance, respectively, where the qualification requirements are generally more severe. The specification scenario in vogue within DoD at the time this book was being written involved preparation of a performance specification for the development process, which drives the design and qualification inspection, followed by preparation of a derivative specification called a detail specification by adding content in a paragraph called design and construction, and changing the flavor elsewhere. The performance specification content should drive qualification, while the detail content drives the acceptance inspection. If the specification includes both performance and detail content, it will have to have associated with it a pair of verification requirements to cover both cases. The verification traceability matrix in this case must have one column for qualification and another for acceptance (and maybe a third permitting different coverage of first article and recurring acceptance inspection, if necessary).

9.1.3 How Does the Detail and Performance Specification Content Differ?

Performance specification content defines the performance requirements and constraints needed to control the design process. These requirements define the problem that must be solved by the design team. They should be solution independent so as not to unnecessarily constrain the solution space or encourage point design solutions. They should define the broadest possible solution space such that the resultant design solution will satisfy its assigned functionality, contributing effectively toward satisfying the system need.

Detail specification content is design dependent and written subsequent to conclusion of the preliminary design. These requirements should identify high-level conditions or measurements rather than defining redundant or detailed requirements. If the product satisfies these requirements, it should indicate the whole item is satisfactory. Table 8, derived from SD-15, Defense Standardization Program Performance Specification Guide, published by the Office of Secretary of Defense, describes the differences in specification content in several major subject areas for the two types indicated. The author very much disagrees with SD-15 and MIL-STD-961E regarding Section 5, especially when this format is used for procurement specifications.

9.1.4 Conversion to Verification Requirements

The content of Section 3 of the detail specification must be converted into verification requirements for Section 4, just as in the case for performance specifications. Here, too, we should write the pairs together to encourage better Section 3 content. It is more difficult to write bad requirements when you must also define how you will prove that the design satisfies them. Also, as in performance specifications, it is possible to include the verification

Table 8 Requirements Comparison

Specification Requirements	Performance Specification	Detail Specification
Section 3 Content		
1. General	States what is required, but not how to do it. Should not limit a contractor to specific materials processes, or parts when Government has quality, reliability, or safety concerns.	Includes "how to" and specific design requirements. Should include as many performance requirements as possible but they should not conflict with detail requirements.
2. Performance	States what the item or system shall do in terms of capacity or function of operation. Upper and/or lower performance characteristics are stated as requirements, not as goals or best efforts.	States how to achieve the performance.
3. Design	Does not apply "how-to" or specific design requirements.	Includes "how-to" and specific design requirements. Often specifies exact parts and components. Routinely states requirements in accordance with specific drawings, showing detail design of a housing, for example.
4. Physical Requirements	Gives specifics only to the extent necessary for interface, interoperability, environment in which item must operate, or human factors. Includes the following as applicable: overall weight and envelope dimension limits; and physical, federal, or industry design standards that must be applied to the design or production of the item. Such requirements should be unique, absolutely necessary for the proper manufacture of the item, and used sparingly. An example would be the need to meet FAA design and production requirements for aircraft components.	Details weight, size, dimensions, etc. for item and component parts. Design specific detail often exceeds what is needed for interface, etc.

5. Interface	Similar for both design and performance specifications. Form and fit requirements are acceptable to ensure interoperability and interchangeability.	Same
6. Material	Leaves specifics to contractor, but may require some material characteristics; e.g., corrosion resistance. Does not state detail requirements except shall specify any item-unique requirements governing the use of material in the design of the item. Such requirements should be unique, critical to the successful use of the item, and kept to a minimum. An example would be the mandated use of an existing military inventory item as a component in the new design.	May require specific material, usually in accordance with a specification or standard.
7. Processes	Few, if any, requirements.	Often specifies the exact processes and procedures and procedures to follow – temperature, time, and other conditions – to achieve a result; for example, tempering, annealing, machining and finishing, welding, and soldering procedures.
8. Parts	Does not require specific parts.	States which fasteners, electronic piece parts, cables, sheet stock, etc., will be used.
9. Construction, Fabrication, and Assembly	Very few requirements.	Describes the steps involved or references procedures which must be followed; also describes how individual components are assembled.
10. Operating Characteristics	Omits, except very general descriptions in some cases.	Specifies in detail how the item shall work.
11. Workmanship	Very few requirements.	Specifies steps or procedures in some cases.

Continued

Table 8 Requirements Comparison—cont'd

Specification Requirements	Performance Specification	Detail Specification
12. Reliability	States reliability in quantitative terms. Must also define the conditions under which the requirements must be met. Minimum values should be stated for each requirement, e.g., mean time between failure, mean time between replacement, etc.	Often achieves reliability by requiring a known reliable design.
13. Maintainability	Specifies quantitative maintainability requirements such as mean and maximum downtime, mean and maximum repair time, mean time between maintenance actions, the ratio of maintenance hours to hours of operation, limits on the number of people and level of skill required for maintenance actions, or maintenance cost per hour of operation. Additionally, existing government and commercial test equipment used in conjunction with the item must be identified. Compatibility between the item and the test equipment must be specified.	Specifies how preventive maintainability requirements shall be met: e.g., specific lubrication procedures to follow in addition to those stated under Performance. Also, often specifies exact designs to accomplish maintenance efforts.
14. Environmental Requirements	Establishes requirements for humidity, temperature, shock, vibration, etc. and requirement to obtain evidence of failure or mechanical damage.	Similar to performance requirements.
Section 4 Content	Must provide both the Government and the contractor with a means for assuring compliance with the specification requirements.	Same as for performance specifications.
1. General	Very similar for both performance and detail. More emphasis on functional. Comparatively more testing for performance in some cases.	Very similar for both performance and detail. Additional emphasis on visual inspection for design in some cases.

2. First Article	Very similar for both performance and detail. However, often greater need for first article inspection because of greater likelihood of innovative approaches.	Very similar for both performance and detail. Possibly less need for first article inspections.
3. Inspection Conditions	Same for both.	
4. Qualification	Same for both.	Same.
Section 5 Content	All detailed packaging requirements should be eliminated from both performance and detail specifications. Packaging information is usually contained in contracts.	

requirements in Section 4 of the detail specification, a special test require-ments document, or in the acceptance test documentation. The author encourages including them in Section 4 of the specification, once again, to encourage good Section 3 content.

Since the Section 3 requirements in the detail specification will be much more physically oriented toward product features than those in the perfor-mance specification, it stands to reason that the verification requirements will be more closely related to specific equipment or software code features and measurements. We may call for specific voltage readings on particular connector pins under prescribed conditions, for example.

9.1.5 Acceptance Test Planning, Procedures, and Results Data Collection

This text links the word *test* to acceptance work while it should be generally referred to as *inspection*, meaning it could be accomplished using any of the five methods described under qualification. It is simply an old habit to con-sider test first. We will discuss nontest methods shortly. The acceptance test plan and procedure for an item should be an expression of the verification requirements. A test condition will commonly call for a series of set-up steps that define the input conditions, followed by one or more measurements that should result. Many of these sequences strung together provide the complete test procedure. This process of product detail specification requirements definition, acceptance requirements definition, and accep-tance test planning forms a sequence of work moving from the general to the specific, from a condition of minimum knowledge toward complete knowledge, which conforms to a natural thinking and learning pattern for people.

The verification requirements give us knowledge of what has to be done and with what accuracy. The test plans give us detailed knowledge of test goals and identify required resources (time, money, people and skills, mate-rial, supporting equipment, and the unit under test) to achieve those goals. The test procedures provide the persons responsible for doing the test work a detailed procedure so as to ensure that the results provide best evidence of the compliance condition. The data collected from the test should reveal that best evidence useful in convincing us and our customer that the product sat-isfies its detail specification content, if in fact it does. This process should not be structured to produce supportive results regardless of the truth, but rather to produce truthful results. If the results yield a conclusion that the product does not comply, then we have the same problem we discussed under qualification except that we are further down the road. The possibilities are the following:

a. If we conclude that the requirements are unnecessarily demanding, we could change the requirements and procedures to reflect a more realistic condition. This may require the cooperation of your customer, of course. This is an easy conclusion to reach because it can solve a potentially costly problem late in the development process, but there is a very thin line between sound, ethical engineering and unethical expediency in this decision. We may have to retest in this case, but if it can be shown analytically that the current design will satisfy the new requirements based on its performance in the failed test, it may not be necessary.

b. Change the design and/or the manufacturing process to satisfy the requirements and retest.

c. Gain authorization from the customer for a waiver allowing delivery of some number of articles, or for a specific period of time, that fail to satisfy the requirements in one or more specific ways. This waiver condition could exist for the whole production run, but more commonly it would extend until a predicted cut-in point, where design changes will have been incorporated that cause the product to satisfy the original requirements.

9.1.6 Associate Contractor Relationships

The verification work across these interfaces for production purposes may entail special matched tooling jigs. For example, the builder of an upper stage of a space transport rocket that interfaces with the payload manufactured by another contractor may be required to fit a special payload tooling jig, supplied by the payload contractor, to the upper stage to verify that when the two come together at the launch site the mechanical interface will be satisfactory. Other tests will have to be accomplished by each contractor, to verify electrical, environmental, and fluid interfaces in accordance with information supplied by the other contractor. Each mission may have some unique aspects, so this interface acceptance process may include generic and mission-peculiar components.

The acceptance requirements for these interfaces should be defined in the interface control document or interface specification and agreed upon by the associates. Each party must then translate those requirements into a valid test plan and procedure appropriate for their side of the interface. Ideally, each contractor should provide their test planning data to the other for review and comment.

9.1.7 Manufacturing or Test and Evaluation Driven Acceptance Testing

In organizations with a weak systems community and a strong manufacturing or test-and-evaluation community, the route to acceptance test planning and detail specifications may be very different from that described up to this point in this chapter. Manufacturing or test and evaluation may take the lead in developing the acceptance test plan prior to the development or in the absence of the development of a detail specification. In this case, the test requirements are defined based on experience with the company's product line rather than written product-specific requirements. Many companies do a credible job in this fashion, but the author does not believe it offers the optimum process. In such companies the detail specification would be created based on the content of the acceptance test plan, and this is backwards. The value of the work accomplished in creating the detail specification in this case is probably not very high.

9.1.8 Information Management

As in qualification verification, there is a substantial need for data in the acceptance testing process. This stream is similar to that for qualification testing, including:

a. a detail (or product) specification containing product requirements, a verification traceability matrix defining methods, level, and reference to the corresponding verification requirements, ideally located in Section 4 or 6 of that same document;

b. a test plan and procedure; and

c. a test report that may include nothing but data sheets.

Just as in qualification testing, all of this documentation can be captured in the tables of a relational database. The data collection process could include inputs directly from the factory floor through integration of the verification database with the production test support equipment.

9.1.9 Coordination between Acceptance Testing and Special Test Equipment

The design of the special test equipment (STE) that will have to be created to support the production line should be based on the acceptance test requirements. As discussed earlier, the acceptance test measurements should be end-test points rather than intermediate points. Therefore, the STE may require additional capability, where the acceptance testing is only a subset. This additional capability would be introduced to permit troubleshooting and fault isolation, given that a particular acceptance reading did not measure up to the required value. STE may also require the capability to provide the right conditions for factory adjustments or calibrations.

9.1.10 Relationship between Technical Data and Acceptance

The acceptance testing process may be very close to what is required in the field to verify that the product is ready for operational use after a prior application. The test procedure would simply be captured in the technical data, with some possible changes reflecting the differences between STE and field support equipment and the hard use the product may receive between test events. The measurement tolerances should reflect a formal tolerance funneling plan, including item and system qualification as well as acceptance test and as many as three levels of field test situations. Tolerance funneling calls for a progressively less demanding tolerance as you move from the factory toward the ultimate user. The purpose is to avoid possible problems with product that fails field testing only to be declared acceptable at depot level (that could be the factory).

The qualification testing should be more severe than acceptance testing and require a tighter tolerance. A unit completing qualification testing generally is not considered acceptable for operational use. That is, it should not normally be delivered to the customer. The acceptance test levels are not intended to damage the item, obviously, in that they are a prerequisite for customer acceptance of the product. All testing subsequent to acceptance should apply the same measurements but accept a widening tolerance, as indicated in Figure 63. In this case the procedure calls for reading voltage X at a test point plus or minus a tolerance. Reading 1, taken at acceptance of the unit, would be acceptable if taken at any level of testing. Reading 2 would be acceptable at field intermediate level and on the borderline for depot test, but it would not be accepted if the unit were shipped to the factory for repair and testing.

A story is offered to dramatize the problems that can go undiscovered. The author investigated several unmanned aircraft crashes while a system engineer in the 1970s at Teledyne Ryan Aeronautical. On one of these, an AQM-34 V electronic warfare unmanned aircraft had been launched from a DC-130 over a desert test and training range in the southwest corner of Arizona. Fifteen seconds after launch it dove directly to the desert floor and buried itself in the desert. When the author reached the crash site, he asked the troops to see if they could find the flight control box in the black hole, because this was the unit that was supposed to have placed the vertical control axis on an altitude control mode 15 seconds after launch.

Shortly, the box was found in surprisingly good condition. The unit was analyzed at the factory, and the altitude transducer opened for inspection.

ACCEPTANCE TASK	ENGINEERING			PRODUCTION	QUALITY
	D&A	SE	T&E		
PREPARE ATP	P		S		
APPROVE ATP		P	R	R	R
IMPLEMENT ATP				P	
ASSURE COMPLIANCE WITH ATP					P

P   PRIMARY RESPONSIBLITY
R   REVIEWER
S   SUPPORTING ROLE
ATP   ACCEPTANCE TASK PLAN/PROCEDURE
D&A   DESIGN AND ANALYSIS
SE   SYSTEM ENGINEERING
T&E   TEST AND EVALUATION

Figure 63 Tolerance funneling.

It was found that there was no solder in one of the three electrical pin pots inside the transducer. The history on this serialized unit revealed that it had been sent back to the transducer manufacturer more than once for a "can-not-duplicate" problem reported in the field. It could not be duplicated in the factory, so it was returned to stock and eventually used in the flight control box in question. Apparently, the wire made sufficient contact with the pin in a static situation to pass all ground tests, but when the vehicle hit the dynamic condition of an air launch pullout with its engine running, the contact of the unsoldered pin failed. The logic of the vertical control axis concluded that the altitude command was zero feet (zero volts) above sea level, and the vehicle's race to achieve that level was interrupted by the desert floor considerably above sea level.

The final answer to this specific problem was provided by vibration testing the flight control box and the uninstalled transducers as part of their acceptance test process. Granted, this is more of a binary condition and tolerance funneling is more properly concerned with a graduated or analog response situation, as noted in Figure 63. But the reader can clearly see from Figure 63 that if the unit under test in the field fails to satisfy the field test tolerance, then it is unlikely that it will pass the factory or depot test. This preserves the integrity of the aggregate test process.

The previous example also highlights another concern in acceptance testing. During qualification testing, it is normal to exhaustively test items for environmental conditions using thermal, vibration, and shock testing apparatus. It is not uncommon to forego this kind of testing when we move to acceptance testing, based on the assumption that the design is sound, having passed qualification. We should eliminate all environmental testing in acceptance only after considering whether there are manufacturing or material faults, like the unsoldered pins in the previous example, that could go undetected in the absence of all of these environmental tests.

9.1.11 Post-Delivery Testing Applications

It may prove useful to rerun the acceptance test on some products periodically subsequent to delivery. If the product may be stored for some time after delivery and before operational employment, there will be some maximum storage time after which certain components must be replaced, adjusted, or serviced. These are often called *time change* items. The design should group all of the time change items into one or some small number of cycles, rather than require a whole string of time change intervals for individual items to reduce the maintenance and logistic response to a minimum.

When one of these time change limits on a particular article is exceeded, the user may be required to make the time change within a specified period of time subsequent to the expiration, or be allowed to make the change whenever the article is pulled from storage in preparation for use. This is primarily a logistics decision unless failure to change out the time change items can result in an unstable or unsafe condition, as in the case of some propellants and explosives. Subsequent to replacement actions, the product should be subjected to some form of testing to verify the continuing ready status of the article. This testing could be essentially the same as the acceptance test or some subset thereof.

9.1.12 Intercontinental Acceptance

Some companies are beginning to apply networking and email concepts to acceptance testing. At one time Compaq Computer developed acceptance test scripts in its facility in Houston, Texas, and these scripts were sent out over its lines to factories in several foreign countries, where they were installed on test hubs and used to test computers coming down the line.

Imagine how these ideas can be applied in the future with commercially available high-speed internet. Currently, a customer feels comfortable when all of the elements of its product come together at their prime contractor and everything is tested together in the same physical space. Creative contractors will find ways to apply data communications to link together the elements at different locations around the world in an acceptance test that is virtually the same as if all of the things were physically together on one factory floor. The payoff will be reduced cost of shipment of material and the time it takes to do it.

This same concept can be applied to the qualification process as well, where computer simulations may interact with software under development and hardware breadboards distributed at development sites located in Cedar Rapids, Los Angeles, Yokohama, Moscow, and Mexico City. Concurrent development will then leap out of its current physical bindings. Hopefully, by then, we will have learned how to get the best out of the human component in this environment.

9.2 NONTEST ITEM ACCEPTANCE METHODS COORDINATION

9.2.1 Organizational Responsibilities

The whole acceptance process could be accomplished by a test and evaluation organization, manufacturing organization, or a quality engineering organization. In many organizations, this work is partitioned up so that two or even three of these organizations are involved. Design engineering could even be involved with some analyses of test results. Mass properties may require access to manufactured products to weigh them. Software acceptance may be accomplished by a software quality assurance (SQA) function independent of the QA organization. These dispersed responsibilities make it difficult to manage the overall acceptance process, assure accountability of all participants, and acquire and retain good records of the results so that they are easily accessible by all.

In a common arrangement, the acceptance test procedures are prepared by engineering; manufacturing is responsible for performing acceptance tests and producing test results. The tests are witnessed and data sheets stamped by QA. Demonstrations may be accomplished by manufacturing or engineering personnel and witnessed by quality. Some tests may require engineering support for analysis of the results. Quality assurance may be called upon to perform independent examinations in addition to their witnessing of the actions of others, which qualify as examinations in themselves.

An organizational arrangement discussed earlier pools QA, test and evaluation, and system engineering into one organization focused on system requirements, risk, validation, verification, integration, and optimization. This organization is made responsible for all verification management actions, whether they be qualification or acceptance, and some implementation actions. This arrangement was conceived as a way of centralizing V&V process control, which is very seldom fulfilled in practice, and of stimulating discussion. Commonly, acceptance test is well developed and implemented, once again, because of a strong and competent test and evaluation organization. The acceptance verification tasks accomplished by other methods are, however, easily lost in the paperwork. Our goal in this chapter is to find ways to bring the other methods of acceptance verification into the light and under the same scrutiny as test.

9.2.2 The Coordination Task

No matter which way your company has chosen to organize to produce product and verify product acceptance, the overall verification program coordination fundamentals cannot be ignored. Figure 64 illustrates one way to establish order in this area through a matrix. All of the acceptance tasks are listed on one axis of the matrix and the responsible organizations

1	Report Introduction
1.1	Purpose
1.2	Overview
1.3	Verification Item Definition
1.4	Item Data Organization by Method and Task
1.5	System Verification Reports Organization
2	Item Qualification Verification Reports
3	Item First Article Acceptance Verification Reports
3.1	Item 1 First Article Acceptance Verification Reports
3.1.1	Item 1 Test Task Reports
3.1.1.1	Item 1 Test Task 1 Report
3.1.1.1.1	Test Data
3.1.1.1.2	Evaluation/Analysis of Test Data
3.1.1.1.3	Conclusions
3.1.2	Item 1 Analysis Task Reports
3.1.3	Item 1 Examination Task Reports
3.1.4	Item 1 Demonstration Task Reports
4	Item Recurring Acceptance Verification Reports
4.1	Differences Between First Article and Recurring Acceptance
4.2	Recurring Acceptance Report Access
5	System Verification Reports

Figure 64 Nontest acceptance verification method responsibilities coordination.

on the other axis. ATP, commonly an acronym for *acceptance test plan*, has the meaning *acceptance task plan/procedure* here, to extend the kinds of tasks to include all four methods that may be applied.

In the matrix, we indicate the kinds of responsibilities required of these organizations for the indicated tasks. One example is shown. We could arrange these responsibilities in many different ways. This figure is not intended to support or encourage the application of serial methods in the context of a strong functional management axis of the matrix. The intent is that the product teams formed on the program would have membership by the indicated specialists (from these functional organizations) for the indicated reasons. In the case of manufacturing and related activities that dominate a program at the point where acceptance work is being accomplished, these teams may be organized around the manufacturing facilities and/or led by manufacturing people. During the time the acceptance task plans/procedures are being developed and approved, the program may still be functioning with engineering dominated teams built around the product at some level of indenture.

9.2.3 Acceptance Task Matrix

Just as in qualification and test acceptance verification, we should have a master list of all acceptance verification tasks that must be applied to the product. This matrix will tell what method is involved and establish responsibility for accomplishing the work described in the acceptance portion of the integrated verification plan. There will be a tendency to focus only on the acceptance test tasks, and we have to consciously consider all of the methods to be applied in the acceptance process.

9.2.4 Examination Cases

9.2.4.1 Quality Acceptance Examinations

Two kinds of quality acceptance examinations exist: (1) witnessing of the actions of others to provide convincing evidence that those actions were in fact accomplished and accomplished correctly in accordance with written procedures, and (2) specific examination actions performed on the product by quality personnel.

In the first case, a quality engineer observes the work performed by others, compares it to a work standard at specific points in the process or in a continuous stream from beginning to end. The quality engineer reaches a conclusion about the correctness of the performance and the recordings of the results throughout the work activity. The action observed could be a test, examination, demonstration, or analysis, although tests are the most commonly witnessed activities. In order for the results to be disbelieved, a conspiracy involving at least two people would have to be in place, a situation generally accepted as unlikely. Some very critical integration actions, especially in space applications, may require a two-man rule to guard against the potential for one person to purposely or inadvertently introduce faults. It may be possible to convince a customer that one of these people is performing a quality function, but the two-man action will likely have to be witnessed by a quality engineer as well.

The quality engineer as a monitor of correct performance of work done by others is often implemented by using acceptance test data sheets, where the person performing the test enters the results of the tests and they are signified on the sheet, as witnessed by the quality engineer by an impression of a uniquely identifiable stamp registered to that inspector or by simply initialing the entries. The resultant data sheet becomes part of the acceptance verification evidence.

The acceptance process for a product may also include direct quality assurance action as well as these kinds of indirect actions. The following are examples of direct actions:

a. At some point in the production process, when a particular feature of the product is best exposed and the earliest in the production process that that feature is in its final, unalterable configuration, a quality engineer may be called upon to make a measurement or observation of that feature against a clearly defined standard. This could be done by unaided visual means, use of a simple tool like a caliper and ruler or gage blocks, application of a tool designed to give a clear pass/fail indication visually or by other means, or comparison of the feature with the related engineering drawings.

b. An inaccessible feature of the product is subjected to an x-ray inspection and observed results compared with a pass/fail criterion.

c. A multiple-ply fiberglass structure is examined by a quality engineer for freedom from voids by striking the surface at various places with a quarter (U.S. 25-cent piece) and listening to the resultant sound. One experienced in this examination technique can detect voids with great reliability at very low examination apparatus cost.

It is granted that the difference between the two kinds of quality assurance described previously is slim, and that they merge together as you close the time between the original manufacturing action and the quality examination. In the first case, the action is being observed in real time. In the latter case, the observation is delayed in time from the action being inspected. Generally, quality examinations should be made in real time, but this is not always possible or, at least, practical. Some examples follow:

a. Work Space Constraint. It may not be possible for two people to locate themselves so that the manufacturing person can do the work while the quality person is observing.

b. Safety Constraint. During the time the action is being performed, there is some danger for the person performing the work, but the article is relatively safe after the act is complete, and the effects of the work can be clearly observed after the fact.

c. Cost Constraint. Each examination costs money, and the more we introduce, the more money they cost. Also, each examination is a potential source of discontinuity in the production process, if it is necessary for manufacturing work on the product to cease during the examination. If the examination drags out in time, manufacturing labor can be wasted as they remain idle. The cost of making sure it is right is reflected in the manufacturing cost and may, in the aggregate, show up as a manufacturing overrun in the extreme case, for which manufacturing must answer. This is a common area of conflict between quality and manufacturing and a sound reason why quality must be independent of the manufacturing management hierarchy.

Where the manufacturing action and examination action are close coupled, it is generally obvious what manufacturing action or actions are covered by the examination. Where some time does elapse between a manufacturing action and the related quality examination, no matter the cause, it should be noted in the quality examination procedures or in some form of map between these data sets precisely to which manufacturing action or actions the examination traces.

Clearly these examination points must be very carefully chosen in the interest of cost and effectiveness. Their location in the production process should be selected for a lot of the same reasons used when deciding where to place equipment test points for troubleshooting or instrumentation pickup points for flight test data collection in a new aircraft. We are, in effect, instrumenting the production process for quality assurance measurements. Ideally, these measurements should be capable of statement in numerical form such that a metric approach can be applied over time or per unit of articles, and these numbers used to chart process integrity using statistical process control techniques.

9.2.4.2 Engineering Participation in Acceptance Examination

The need for routine participation of design engineers in the acceptance examination process should be viewed as a symptom of failure in the development process. A example of where this can happen is the development of the engineering for an item using sketches rather than formal engineering drawings, in the interest of reduced cost in the early program steps, with the intent to upgrade the engineering between completion of qualification and high-rate production. Should it later develop that there are no funds to upgrade the engineering, but the customer is enthusiastically encouraging you to proceed into production with full appreciation of the potential risk, it may be possible for very qualified manufacturing people to build or assemble the items incorrectly just because of ambiguities in the engineering. Without engineering participation, this kind of problem can lead to unexplained production line problems that are finally traced to a particular person being out sick, retired, or promoted. That person had determined how to interpret the engineering and that knowledge did not get passed along through any written media so that the replacement could unfailingly do the work correctly.

Engineers could be brought into the examination process, but it should be recognized as a short-term solution to a problem driven by poor engineering documentation. As these problems are exposed, engineering should work those off as budget permits such that, at some time in the future, manufacturing and quality people can properly do their jobs unaided by direct engineering support.

9.2.4.3 Software Acceptance Examination

Computer software is a very different entity from hardware, and its acceptance is handled in a very different way. If we thoroughly test a software entity in its qualification process and can conclude that it is satisfactory for the intended application, it would seem that we have but to copy it as many times as we wish to manufacture it. Each copy of the software should have precisely the same content as the original proven in qualification. The reliability of the copying process is very high, and most people would accept that the copied article is identical to the original. In order to make absolutely certain that this is the case, the copied article could be subjected to a comparison with the original. Any failures to compare perfectly should be a cause for concern, in that the two files should be identical. Alternatively, the copied article could be subjected to a checksum routine and the results compared with the original checksum.

The knowing customer's concern for software acceptance might be more closely related to uncertainties in the integrity of the developer's process than the simple correlation between master and delivered product. The question becomes, "Is the current software master the same code that passed qualification?" This is especially a concern after one or more changes to the product subsequent to its initial development, qualification, and distribution. These concerns can best be answered by the developer applying sound software configuration management principles in the context of a well-managed software library with access rigorously controlled and histories well kept. There should be a believable, traceable history of the product configuration and actions taken to the master during its life cycle.

9.2.5 Demonstration Cases

9.2.5.1 Logistics Demonstrations

One of the key elements in many systems is the technical data supplied to the customer to support their operation and maintenance of the system. This data is generally accepted as a lot where it is delivered in paper document form. The customer may require 100 copies of each manual and accept them as a lot based on the technical data having been verified and all problems discovered during verification having been resolved. The words *validation* and *verification* are used a little differently here. The validation process often occurs at the contractor's facility, and the procedures called for in the data are accomplished by contractor people, perhaps using qualification and system test articles. The verification work is then done at a customer facility when sufficient product has been delivered in the final configuration subsequent to FCA/PCA to support the work.

It is entirely possible and very common that the verification work will uncover problems not observed during validation. One of the reasons for this is that the people doing the work in validation are engineers very familiar with the product as a result of their work over a period of what may have been several years. These people may see things in the procedures that are not there, while someone from the customer's ranks during verification will not be able to complete the same procedure because the equivalent knowledge level has not been achieved. The data may have to be changed in verification to make it more specific and complete so that anyone who has completed some prescribed level of training can complete the work.

The author discovered this lesson the hard way as a field engineer with SAC at Ben Hoa Air Base in South Vietnam. He had deployed in support of a new model of the Ryan series of unmanned aircraft called a Ryan model 147H. During flight test he had found a problem in the compartment pressurization test procedure in technical data and rewrote the whole procedure. Apparently, it was never validated other than mentally or analytically, due to the pressure of many more serious problems. Also, it was never verified at Davis-Monthan Air Force Base either, where the USAF unmanned aircraft group was headquartered and from which the vehicle had been deployed. During the buildup of the first vehicle after deployment, one evening one of the troops came into the tech rep trailer and very irately asked who the idiot was who had written the pressurization test procedure. The author owned up that he was the idiot when he found that, in order to attach the compartment pressurization test source pressure hose, one of the compartment doors had to be open, violating the pressure integrity of the compartments. It took the rest of the night to rewrite the procedure and run the test, finally verifying it. (It was a little more embarrassing because the author's boss had deployed with the vehicle as well, resulting in the author being temporarily fired again.) Luckily, the procedure was made to work before the next flight out to the USA was scheduled.

9.2.5.2 Flight Demonstration

When an aircraft manufacturer ships its product, the aircraft will be flown to its destination. This could be considered a final demonstration of the airworthiness of the aircraft. Each aircraft would previously have had to complete a series of flights to prove this, of course. The first flight of the model will have to go through a complete flight test and FAA certification process involving instrumented flights, but the routine delivery of aircraft 56 will not have to undergo this whole battery of tests. The author would consider these kinds of flights demonstrations rather than tests since they do not involve extensive instrumentation, but rather just the product being used in accordance with a predetermined plan. The pilot is in effect demonstrating the capabilities of the aircraft and that they match the required capabilities.

9.2.6 Analysis Cases

In general, test results are much preferred to analysis results for customer acceptance. However, some of the tests that are applied may produce results that are not immediately obvious as to whether or not the product satisfies the requirements. In these cases, analysis of the data may be required. The analysis tasks, just like the verification tasks mapping to the other methods, should be listed in our task matrix and identified in terms of who or what organization is responsible. The analysis task should be described in the acceptance analysis portion of our integrated verification plan.

Our work scheduling for the manufacturing process should identify these acceptance analyses causing the assigned person(s) to accomplish them, producing a conclusion documented in some way for inclusion in the acceptance item record.

9.3 PRODUCT ACCEPTANCE VERIFICATION REPORTING

9.3.1 Program Type

We can group all programs into three types, which should have significantly different acceptance processes applied. On programs characterized by a high-rate production process producing many articles, there is generally sufficient funding available to provide a dedicated set of items for qualification. A second program type produces a few articles of considerable cost, referred to by the author as a low-volume, high-dollar program. On these programs you can't afford to dedicate a set of articles for engineering testing and qualification. A common approach in this case is to send the first article through a reduced stress qualification process, followed by a refurbishment to restore full life capability to the article. Subsequent articles are subjected to acceptance levels only and the first article becomes the last item through acceptance after it has been refurbished. This item may become a spare on the program with the hope that it need never be used in an operational sense. Finally, a program may produce only a single article, such as a manufacturing plant, a chemical processing facility, or a ride at an entertainment park. Obviously, a dedicated article is out of the question. The only sensible thing to do is to combine a nondestructive qualification process with acceptance.

In this chapter, we will proceed based on the high-rate production case. The other two program types commonly will call for a merged set of planning, procedures, and reports driven by a one-part series of specifications.

9.3.2 The Two Steps of Acceptance

Acceptance actually commonly takes place in two steps on a high-rate production program. The first article of a particular item may be subjected to special first-article verification tasks, leading to an audit of the reports produced based on these tasks. This audit is commonly called a physical configuration audit, as previously defined. Subsequently, every article of that item will be subjected to a recurring acceptance process. The evidence from the first article series must be retained for audit. The recurring acceptance process will produce a lot of data. This book encourages capturing the first-article reports in the integrated verification data report (Section 3), providing all of the report documentation needed for PCA.

9.3.3 Significant Differences

The item qualification process is applied to one item generally, though it may be applied in an ascending product entity structure scope, resulting in multiple items actually being tested. In so doing, we commonly conclude qualification with used-up items mostly due to an environmental requirements proof process and a search for margins achieved. The items subjected to qualification cannot be depended upon to have any residual life subsequent to qualification. Acceptance is commonly performed on every article of every item produced, and these articles must have full life expectancy subsequent to that action. It is therefore uncommon to subject an item to environmental stimulus during acceptance testing. The understanding in acceptance is that you have already proven, through qualification, that the design is sound relative to the full range of requirements in the development specification. In acceptance you are interested in whether that particular article is acceptable for delivery to the customer.

There is also a significant difference in the driving requirements document for the two kinds of item verification. The design and qualification process is driven by the Part I, development, or performance specification, which should contain nothing but design-independent content. The item acceptance process is driven by the content of the Part II, product, or detail specification, which contains design-dependent content. The detail specification should drive the development of the acceptance verification plans and procedures.

The qualification process is run on articles that have commonly been manufactured by highly skilled technicians and mechanics in engineering laboratories while trying to reflect final manufacturing plans, whereas the articles upon which acceptance is performed are manufactured on a real production line.

9.3.4 Reporting Structures

The same reporting alternatives exist for acceptance verification evidence as exposed in Chapter 8 for qualification. We can prepare independent reports for each acceptance task, capture the results in the task verification report, or use database content directly. The two kinds of acceptance are split into Sections 3 and 4, with Section 3 containing all of the first-article inspection report data and Section 4 containing reference to the voluminous data that will be collected over the production run of the system articles.

9.3.5 First-Article Acceptance

Production acceptance inspections commonly collect numerical values on checklist-like forms that are then stamped or signed by a quality engineer

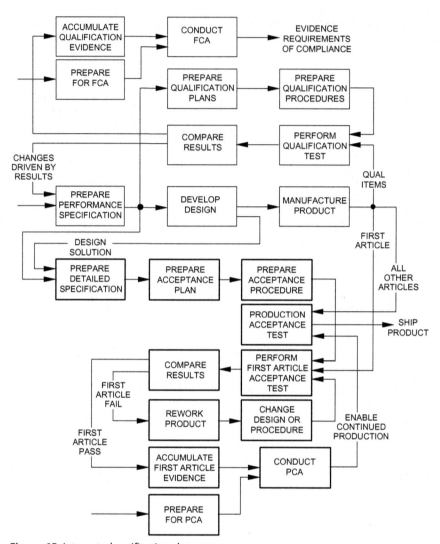

Figure 65 Integrated verification data report structure.

to signify that the work was accomplished in accordance with appropriate procedures. These can be collected during first-article acceptance and included in the integrated verification data report (IVDR) as noted in Figure 65. Ideally, all of this data would be captured in a computer database, with one view being Section 3 of the IVDR.

Section 3 of the IVDR should accumulate all of the product-oriented acceptance data required to support the item PCA. Most of this data will generally be either test or examination data for acceptance. In addition to this data, the PCA auditors may require access to manufacturing planning and engineering drawings.

9.3.6 Recurring Acceptance Verification

For a product that will be manufactured in the hundreds or thousands of articles, the amount of acceptance data may be voluminous. This volume can be considerably constrained by focusing on capture of only the numerical values for specific data entities and the ultimate visual inspection results in terms of pass or fail, in the context of a predefined data collection sheet or a series of computer screen windows.

Acceptance data collection is not just a way to fill available computer storage media with data. It offers a way to maintain control of the manufacturing process. These measurements that are made on the product provide information about the product, yes, but they also tell us the condition of the manufacturing process and resources used in that process. If we see an adverse trend in product measurements, it is telling us something about the process that needs attention. Thus, the data that we collect during the acceptance process should have dual applications recognized in the development of the detail specifications.

9.4 PRODUCT VERIFICATION MANAGEMENT AND AUDIT

9.4.1 The Second Stage of Verification

The qualification process is intended to prove that the design synthesized from the item's development requirements satisfies those requirements. The development requirements, captured in a development, Part I, or performance specification, were defined prior to the design effort so as to be design independent, permitting the person or team responsible for synthesis to consider alternative competing design solutions and make a selection of the best overall solution through a trade study or other organized alternative evaluation process. This series of verification actions is only accomplished once to prove that the design is sound, and the evidence supporting this conclusion is audited on a Department of Defense program in what is called a functional configuration audit (FCA). The tests accomplished during this work commonly stress the item beyond its capabilities, to expose the extent of capabilities and extent of margins available in normal operation of the item and to validate the models used to create the designs. These stresses reduce the life of the unit and may even make an item immediately unusable as a result of a successful qualification series. Therefore, these same inspections cannot be applied to every item produced for delivery, since the delivered items would not have their planned service life at time of delivery.

What form of inspection might then be appropriate for determining whether or not a specific article of the product is acceptable for delivery? This represents the second stage of verification, to prove that the product is a good manufacturing replication of the design that was proven sound through qualification. The acceptance process is accomplished on every article produced, on every product lot, or through some form of sampling of the production output. The process leading to the first manufactured item is subject to audit on Department of Defense programs in what is called a physical configuration audit (PCA). Some customers may refer to this as a first-article inspection process. The word *inspection* is used in this chapter to denote any of the five methods already discussed: test, analysis, examination, demonstration, and special.

9.4.2 The Beginning of Acceptance Verification

The acceptance process should have its beginning in the content of the Part II, product, or detail specifications. This kind of specification is created based on a particular design solution and gives requirements for the physical product. This may include specific voltage measurements at particular test points under conditions defined in terms of a set of inputs, for example. Measurements will have to be made during a test and compared with requirements. As an alternative to two-part specifications, you may choose to use one-part specifications. In this case the specification may have to have a dual set of values or tolerances, one for qualification and another for acceptance. Subsequent to testing, the item must have a full lifetime before it. In either specification case, the beginning of the acceptance process should, like qualification, be in specification content.

Unfortunately, this is not the case in many organizations. In many organizations, the acceptance test procedures are prepared independently of the specifications based on the experience of Test and Evaluation and Quality Assurance department personnel and their observations of the design. Subsequently, if a Part II specification is required, the test procedures are used as the basis for it. This is exactly backwards. It is analogous to accomplishing the design of a product and then developing the specification. The Part I specification is the basis for the design of the product and the qualification process, while the Part II specification is the basis of the design of the acceptance process. If you first design the acceptance test process, then it makes little sense to create the requirements for the process subsequently. So, a fundamental element of acceptance management should be to insist on acceptance verification requirements as a prerequisite to acceptance planning activity.

Figure 66 illustrates the overall qualification and acceptance processes. In this chapter we will focus on the blocks with the heavy border beginning with the development of the detail (Part II or product) specification. This process includes an audit of the acceptance evidence called, in DoD parlance, a physical configuration audit, but the process continues as long as the product remains in production.

9.4.3 The Basis of Acceptance

When dealing with a large customer such as DoD or NASA, the acceptance process should be driven by what provides the customer with convincing evidence that each article or lot of the product is what they want, and what they paid for. This evidence pool should also be minimized based on end-to-end parameters, system-level parameters, and parameters that provide a lot of insight into the condition of the product. So, of all of the things that can be inspected on a product, we have to be selective and not base our evaluation on the possible, but rather on the minimum set of inspections that comprehensively disclose the quality and condition of the product.

Some people feel that the content of the detail (Part II or product) specification should be determined prior to the design. That, frankly, is impossible. As shown in Figure 66, we develop the performance (Part I or development) specification as a prerequisite to the design. Its content should be design independent. The detail specification must contain design-specific information because it is focused on actual observations of the product features, voltage measurements, physical dimensions, extent of motions, and so forth. Therefore, the content of detail specifications must be created subsequent to the development of at least the preliminary design, with a possible update after the detailed design has been formulated.

9.4.4 Acceptance Documentation

The suite of documentation needed as a basis for a sound item acceptance inspection includes the following:

a. Detail Specification Table 8 suggests valid content of the detail specification relative to a performance specification. This content could be integrated into the same specification used for development. These specifications should be prepared during the detailed design phase.

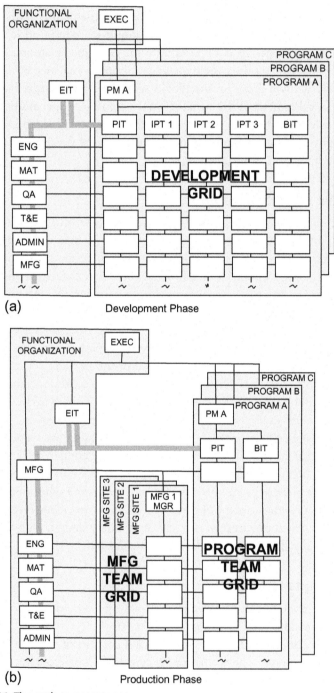

(a)   Development Phase

(b)   Production Phase

Figure 66 The path to acceptance.

b. Quality Inspection Plan Quality Assurance should fashion an inspection plan based on the evolving design concept that identifies specific areas that should be inspected and the most appropriate times and conditions for those inspections in the evolving manufacturing plan. The goals of each inspection should be clearly defined.

c. Acceptance Test Plan Assuming that the product has some functional capabilities rather than simply a physical existence, a test of some kind will be appropriate using instrumentation of some kind to convert reactions to a form suitable for human viewing. Each of the tests to be performed should be identified and defined in terms of goals, resources, budget, and schedule needs.

d. Acceptance Procedures Each inspection identified in the plan should have developed a procedure telling how to accomplish the work. This information may be integrated into a combined plan and procedure also including all other verification work, but this is not the industry norm.

e. Acceptance Reports The acceptance reports may be as simple as sheets for recording values observed in tests and quality examinations or entail computer-recorded reading that can be reviewed on the screen of a computer terminal. The results should be collected for each article. The customer may require delivery of this data or require that the contractor maintain this data on file for a minimum period of time.

The program schedule should identify these documents and when they should be released. Responsible managers should then track the development of these documents and apply management skill as necessary to encourage on-time delivery of quality documents. Each document should pass through some form of review and approval.

9.4.5 Management of the Work

All of the inspection work should appear on program planning and schedules for each article. Management should have available to them cost schedule control (earned value) system and manufacturing tracking data, indicating progress for each article or lot through the manufacturing process. As impediments are exposed, alternative corrective actions should be evaluated and a preferred approach implemented to encourage continued progress on planned work.

The aggregate of the work that must be managed in the acceptance process falls into two major components. First, the products are examined at various stages of the production process within the producing company and at all of its suppliers. In these examinations, quality assurance persons determine whether or not required manufacturing steps were performed and performed correctly to good result. If everything is in order in the traditional paper-oriented process, the QA person signs their name on or stamps a document to indicate acceptability. The other major inspection is in the form of acceptance testing, where functional capabilities are subjected to input stimulus combinations and corresponding outputs compared with a standard. Commonly, checklists are used to record the results. The test technician records values observed, and a QA inspector initials or signs the list to indicate compliance with written procedures, to good results.

Management must encourage the development of clear manufacturing and quality planning data as well as a sound achievable manufacturing schedule that defines the work that must be done and when it must be done, including the quality inspections and the acceptance tests. These activities have to be integrated into the manufacturing process and coordinated between the several parties. In some organizations, manufacturing looks upon the inspection process as an impediment to their progress, and this wrong-headed attitude can be encouraged, where an attempt is made to layer the inspection process onto a previously crafted manufacturing plan that did not account for the time and cost of the inspection process.

The reader should know what is coming next at this point. Yes, we need a team effort. Here is another situation where several specialized persons must cooperate to achieve a goal that none of them can achieve individually. The overall manufacturing and inspection process (the word used in the broadest sense to include test) should be constructed by a team of people involving manufacturing, quality, material, tooling, and engineering. Ideally, all of these people would have participated in the concurrent definition of mutually consistent product and process requirements followed by concurrent development of mutually consistent product and process designs. All of the planning data would have been assembled during the latter. Then, the management process is a matter of implementing the plans in accordance with planned budget and schedule constraints and adjusting for changing conditions and problems.

The teams effective in the production phase ought to be restructured from the development period, as illustrated in Figure 67. Figure 67a shows the suggested structure in development, with several integrated product and process teams (IPPT) on each development program, along with a program integration team (PIT) and a business integration team (BIT).

An enterprise integration team reporting to the executive (by whatever name) acts to optimize at the enterprise level, providing the best balance of

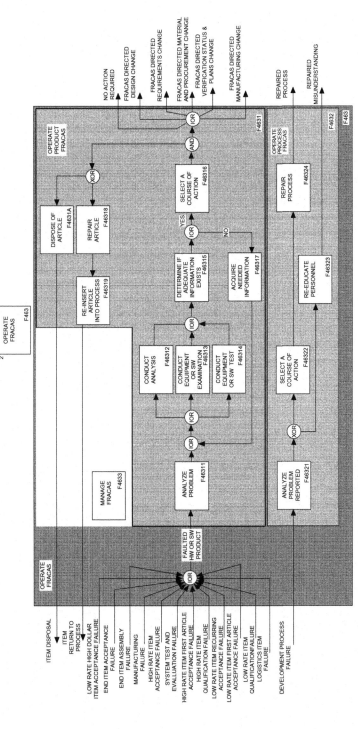

Figure 67 Development to production organizational transform.

enterprise resources to the programs, based on their contractual obligations, available enterprise resources, and profit opportunities. This integration work is denoted by the heavy concurrent development bond joining the several activities. The enterprise integration team (EIT) does integration work across the several programs through the PITs, and the PIT on each program carries on the integration work within the program. During development, the PIT acts as the system agent across the teams, optimizing at the program product and process level. As the product maturity reaches a production-ready state, the IPPTs can collapse into the PIT with continued BIT (contracts, legal, etc.) program support. Production is managed through manufacturing facility managers under the direction of an enterprise functional manufacturing manager, with the EIT optimizing for balanced support to satisfy program commitments. The PIT remains available for engineering in support of production changes and engineering change proposals. PIT specialists support facility teams as required by those teams.

In this arrangement, it is a program responsibility to develop suitable plans and procedures to accomplish acceptance examinations and tests in the manufacturing environment. The production responsibility for particular items is assigned to particular manufacturing facilities and vendors, which are all treated as vendors by the program. These facilities must carry out the planned acceptance work as covered in the approved plans and procedures.

9.4.6 FRACAS

9.4.6.1 Acceptance Ethics

A key element of the production process is a means to acquire information on failures to satisfy acceptance criteria. When items fail to satisfy inspection requirements, the item should not be passed through to the customer without some kind of corrective action. A few companies have gotten themselves in great difficulty with DoD in the past by treating failed items as having passed inspection requirements. In some cases, the tests called for were not even run, in the interest of achieving schedule requirements. In other cases, failures were not reported and items shipped. Clearly, it is necessary to engage in deceit, to put it mildly, in order to perform in this fashion.

Management must make it clear to the work force that integrity of acceptance data and resultant decisions is important to the company, as well as to the people who are doing the work. A company that falsifies records not only damages its customers and its own reputation in time, it also damages those who work there and are required to perform in this fashion. An ethical person simply will not work in this environment. So, either a person's ethics are attacked, or the person replaced with someone who does not have a problem with this arrangement. Clearly, the company is a loser in either case. The moral fiber of the organization has been damaged.

The two causes for this kind of behavior are schedule and budget difficulty. Management persons throughout the chain dealing with acceptance should be alert for those doing the work coming to believe that delivery on time is more important than the company's good name. Unfortunately, this problem generally arises from management encouragement (intentional or otherwise), rather than grass roots creativity.

9.4.6.2 FRACAS Implementation

The organization needs an effective failure reporting and corrective action system (FRACAS) to provide management with accurate information on failures to satisfy acceptance requirements. Figure 68 illustrates a representative system for discussion. Failures can occur, of course, anywhere in the production stream, and the acceptance examination process should detect first that a failure has occurred and second what item or items have been affected by that failure. So, the first step in the FRACAS process is failure detection, whereupon it must be reported in some formal way.

A failure can be caused by a number of different problems, and we should first determine the root cause before settling upon a corrective course of action. We should no more decide a corrective action without understanding the problem here than we should seek to develop a design solution to a problem that we do not yet understand. A fishbone, cause and effect, or Ishikawa diagram is helpful in exhaustively searching for alternative causes. A Pareto diagram may be useful in prioritizing the corrective actions planned. Figure 69 illustrates these diagrammatic treatments, and the reader is encouraged to consult a book on process reengineering, total quality management (TQM), or continuous process improvement for details.

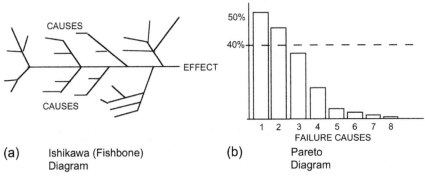

(a)   Ishikawa (Fishbone)
       Diagram

(b)   Pareto
       Diagram

Figure 68 FRACAS process.

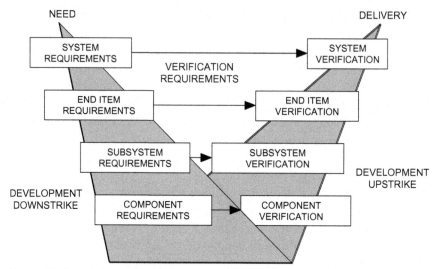

Figure 69 Supporting techniques.

Briefly, the Ishikawa diagram offers a disciplined approach for identifying the effect (failure) one wishes to investigate, and possible causes for that problem. The Pareto diagram shows us that a relatively few probable causes generally drive the negative effect we wish to avoid. Because we seldom have all of the resources we wish were available, we are often forced to prioritize our actions to those areas that will produce the biggest positive effect.

Figure 67 suggests that there must be only a single cause of a failure, and that is not always the case. It may be possible to correct for a given failure in a number of ways. It may be possible to implement a design change, a manufacturing process change, or a procedural change to correct for an observed failure or some combination of these. Our goal should be to select the best corrective action in the aggregate, generally in terms of life-cycle cost effects, but on some programs and in some phases near-term cost may be more important at the time.

9.4.7 Physical Configuration Audit

The PCA provides a way to reach a conclusion about the quality of the manufacturing process as a prerequisite to committing to high-rate production using that process. If there are problems in that process that have not yet been exposed, it is better to expose them earlier than later. Since PCA is performed on the first production article, one cannot get any closer to the beginning of production. It should be pointed out that a sound development process up to the point of entering production of the first article will have a positive effect on reducing the risk that problems will occur in the production process. Another view of this translation was provided by W. J. Willoughby in his transition from development to production templates, documented in NAVSO P-6071 and other places. Appendix A6 offers a physical configuration audit report template.

PCA attracts all of the evidence collected about the production of the first article into one place, where it can be audited by a team of customer and contractor people against the requirements. The result desired by the contractor is that the evidence will be convincing that the product satisfies the detail requirements, conforms to the engineering drawings and lists, and was manufactured in accordance with the approved planning data.

9.4.7.1 PCA Planning and Preparation

As in the FCA, the contractor should develop an agenda subject to customer approval, and prepare and organize the materials to be used at the audit. Because the audit covers a physical reality rather than the intellectual entity audited at FCA, you will have to deal with things as well as paper in the audit. An audit space close to the production area will encourage the most efficient use of time during the audit. The first article should be made available in this space for direct visible examination. All of the acceptance report data should be available in paper form or via computer terminals, depending on the format of this data. The exact arrangements that the contractor makes should be based on the mode of audit selected by the customer.

9.4.7.2 PCA Implementation

The customer may require any one of three different modes for audit presentation. The customer preference should be known to the contractor at the same time that the agenda is crafted, because the different modes encourage very different audit environments. The contractor should also develop and reach agreement with the customer on the criteria for successful completion of the audit, which includes clear exit criteria.

9.4.7.2.1 The Physically Oriented Audit

One of the three modes involves making the product available for customer examination and comparison with the engineering drawings. A U.S. Air Force customer may populate the audit with several staff noncommissioned officers with the right technical skills for the product line. They disassemble the product and compare the parts with the engineering drawings, confirming that the product satisfies its engineering. They may also conduct or witness functional tests on the product to ensure that it satisfies test requirements.

9.4.7.2.2 The Paper-Dominated Audit

The acceptance process generates a considerable amount of evidence in paper or computer media. If the customer elects to apply this mode of audit, a totally different kind of people will be attracted to the audit. They will be people with good research and analytical skills rather than good practical skills. They will review all of the evidence in one of several patterns to assure themselves that the product accurately reflects its controlling documentation. The auditors may selectively evaluate the evidence, looking for inconsistencies. If they find one, they will search deeper; if more are uncovered, they may audit the whole set of evidence. If, on the other hand, no problems are found in the first layer, the audit terminates with a good report. Alternatively, the audit may focus on only one aspect of the evidence, such as qual test reports, using the progressive technique just discussed. A third approach is to simply plan to audit all of the evidence in some particular pattern.

9.4.7.2.3 Combined Mode Audit

A customer could bring in both populations just discussed and perform both kinds of audit. This audit could be coordinated between the two groups to apply a progressive approach, or the plan could be to audit all data.

9.4.7.3 Post-PCA Activity

The results of the PCA should be documented in minutes containing identification of any remaining problems and a timetable for resolving them. The precise mechanism for final closure should be included. These actions may be fairly complex, entailing changes in specifications, changes in design, changes in acceptance test plans and procedures, and changes in manufacturing and quality planning. It will be necessary to track all of these changes while they are in work and coordinate the package for integration into subsequent production articles. Some of the changes may not be possible in time to influence near-term production, unless production is halted for an extended period. The adverse effects of this on continuity of the program team is generally so negative that the decision will be made to continue production with one or more deviations accepted by the customer for a brief period of time until the changes can be fully integrated into the design and planning data. Those units produced under a deviation may have to be modified in the field subsequent to the incorporation of the complete

fix in the production process. If so, modification kits will have to be produced, proofed, and installed in the field to bring these units up to the final
configuration. If the changes are sufficiently extensive, it may even be necessary to run a delta PCA on the first article through the upgraded
production line.

9.4.8 Software Acceptance

Software follows essentially the same acceptance sequence described earlier.
The fundamental difference is that software does not have a physical existence except to the extent that it resides on a medium that will be delivered.
The software will be delivered on tape, disk, or other recording media, and it
is this media that must be accepted. Once the computer software has been
proven to accomplish the planned functionality, the acceptance test may be
as simple as a checksum or as complex as a special test involving operation of
the software in its planned environment with actual interfaces and human
operators at work in preplanned situations similar to the anticipated operating environment.

CHAPTER 10

System Test and Evaluation Verification

10.1 SYSTEM VERIFICATION PLANNING

10.1.1 The Beginning

System verification planning should start all the way back with the development of the system specification. The V diagram illustrates this very well. As noted in Figure 70, the content of the system specification should drive system test requirements crafted at the same time as the specification is written. Ideally, these requirements will be included in the verification section of the specification (4 in a DoD specification). Some organizations place these requirements elsewhere or fail to go through this step, and the result is a poorer specification and unnecessary program risk. If you have to write the verification requirement as the product requirements are being written, you will write better product requirements. There will not be any requirements statements like "shall work well and last a long time" in the specification, because one cannot craft a way to test or analyze whether or not the final product satisfies this kind of requirement, thus encouraging the writer to realize that they cannot write the verification requirement because they have written an unverifiable product requirement.

All of those system requirements mapped to test in the system specification verification traceability matrix should come to rest in the hands of the person responsible for creating the system test plan. These may be augmented by verification requirements, or verification strings in the context of top-tier item verification work, promoted from those items to parent-level (system) verification. As described for items, the system test planning process must determine how to most affordably and most effectively determine the extent to which the system satisfies these requirements.

The first step is to synthesize the test requirements into a set of test tasks. On a simple product, this may only require a single test task. On a new aircraft, this can entail many flights over a long period of time, preceded by many ground tests and taxi tests. The first flight might be a takeoff with the landing gear pinned, followed by an immediate landing, followed by a

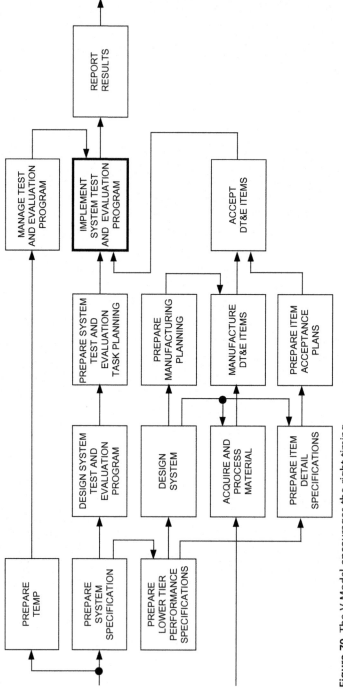

Figure 70 The V Model encourages the right timing.

takeoff and several touch-and-go landings with all flights within visible distance from the flight test base or under surveillance from chase planes. Based on the results of these tests and analysis of any instrumentation data recorded during them, engineering should provide guidance on subsequently planned flights. It may be necessary to alter initial flight plans, restricting certain flight regimes or loading conditions or moving them into later flights until data can be further analyzed.

Generally, it is sound practice to test from the simple toward the complex, from the small toward the large, from the specific toward the general, from the safe toward the more hazardous, and from the easy toward the more difficult. Our system testing should progressively build confidence in future testing by proving the soundness of increasingly larger portions of the design. This approach essentially involves using an increasingly larger portion of the product under test as part of the test apparatus, closing in on the most difficult portions of the system to reach conclusions about.

It should be noted that there is a special kind of verification plan that is crafted very early in a program, often referred to as a test and evaluation (T&E) master plan (TEMP) or integrated test and evaluation plan (ITEP), which is more closely related to program management of the T&E process than the details of verification work to be done.

10.1.2 System-Level Testing and Analysis Implementation
10.1.2.1 Product Cases
The nature of the system testing required during development is a function of the product. An aircraft, helicopter, missile, or remotely piloted aircraft will likely have to undergo some form of flight test to prove that it can actually fly under all required conditions and achieve mission success. A ground vehicle will have to undergo some form of ground testing. A tank, truck, or other kind of vehicle should be required to perform over real or specially prepared terrain of the kind covered in the system specification while verifying, in a military product, offensive and defensive capabilities. In all cases, the end items subjected to system test should be composed of components, subsystems, and systems that have previously been qualified for the application.

Because we have tested all of the parts, there may be a temptation to conclude that it is not necessary to test the whole. This conclusion is sometimes correct, but more often is not. The problem with complex systems is that it is very hard to intellectualize everything. This is true especially in space systems. Many readers will recall the problem with the Hubble Space

Telescope, which could not focus properly when deployed because the optics were not correctly configured. Cost, schedule, and security problems during development with respect to the use of facilities capable of adequately testing the telescope resulted in the rationalization that major elements tested separately would work properly together. Because the telescope was placed in a low Earth orbit, it was possible to modify it while in orbit using the Space Shuttle, but only at tremendous cost and embarrassment.

Space systems are commonly tested throughout the product hierarchy, starting at the bottom and working toward the whole, up through each family structure. This is true during development to qualify the system as well as when assembling a particular deliverable end item for acceptance. The logic for this is pretty simple. Shuttle repair of the Hubble notwithstanding, it is not possible to routinely solve design problems that were not observed in testing during development after launch. We simply do not have the same flexibility in space systems that we have in ground, aeronautical, or ocean-going systems, systems with which we can interact with relative ease after they have departed their manufacturing site but still only in combination with considerable cost.

The intensity with which we will choose to pursue system testing is a function of the planned use of the product, its complexity, and available resources. This should not be a matter of chance or personal preference. One of the common failures in system test planning is to select those things that can be tested based on the current design. System testing should be based on system requirements. We should be trying to prove that the system performs as required in its system specification within the context of prior successful component and end item test results. During the design development, based on specification content, we should be concurrently designing the system test. Where it is very difficult to accomplish a particular verification action at the system level, and it cannot be accomplished at a lower level, we should consider a design change to make it possible. This is, of course, the value of concurrent development. If we can catch these problems early enough, it is economically feasible to develop the best possible system that is also consistent with the right test, manufacturing, logistics, and use processes.

There are relatively few good written sources for the testing of purely hardware systems or the hardware elements of grand systems. Those that the author has found useful are military standards. One of these offers a good superset for testing guidance within which most test situations will be a subset. This is MIL-STD-1540C, Test Requirements for Space Vehicles.

MIL-STD-810D, Environmental Test Methods and Engineering Guidelines, and MIL-STD-1441, Environmental Compatibility Requirements for Space Systems, also provide useful content relative to environmental testing, a significant portion of the test problem for systems and lower tier elements.

A good deal of good literature is available on the testing of computer software, in the form of books written by Boris Beizer and William Perry, to name two respected authors in this field. Software cannot, of course, form a complete system, in that it requires a medium within which to run, a computer of some kind. Given that the machine, its compiler, and the selected language are fully compatible, the software can be treated as an isolated system, however. The software must be thoroughly tested internally and eventually integrated with the hardware system within which the computer operates.

Grand systems, composed of complex arrangements of hardware, software, and people (sometimes referred to by the troubling term *peopleware*) performing procedures based on observations and circumstances, involve very complex system testing demands because it is so difficult to intellectualize all of the possibilities. We are commonly left with few alternatives but to select some small finite number of alternatives from the nearly infinite possibilities and extrapolate across these few test cases. Given that the hardware, software, and procedural elements are each examined carefully throughout their range of variation, there is a reasonable expectation that system testing, so constituted, will be effective in proving compliance with system requirements. Whether or not we will succeed in uncovering all of the possible fault modes undiscovered analytically through failure modes effects and criticality analysis (FMECA) is another matter. In a complex system, we will be hard pressed to imagine all of the strings of activity that can occur in actual operation, especially where the user is intent on getting every possible scintilla of capability out of the system and is not averse to risk-taking to satisfy some borderline operational opportunity.

10.1.2.2 System Test Categories
The Department of Defense (DoD) recognizes several categories of system test that can also be applied to a broader range of customers. The category boundaries are based on who does the work and the conditions in effect rather than on military parameters. In all of these cases, the understanding is that the components and end items that comprise the system have been qualified as a prerequisite to system test.

10.1.2.2.1 Development Test and Evaluation

Development test and evaluation (DT&E) is accomplished by the contractor in accordance with a test plan that focuses on engineering considerations under the contract covering the development of the system. It should be driven by the content of the system specification, and the testing accomplished should produce convincing evidence that the system to be delivered does satisfy the requirements in that specification. This testing is developed by engineers of the producing enterprise to scientifically develop evidence of compliance or failure to do so. Appendix A7 offers a DT&E plan template and Appendix A8 offers a DT&E report template.

This is inherently dangerous work because it commonly entails large masses moving in space and time, often involving hazardous materials, especially with military systems. Safety must, therefore, be an integral part of all system test and evaluation planning. A progressive approach is a good idea, where one starts with static and low dynamic stress tasks, and given success, more dynamic situations are exercised.

System testing often cannot be accomplished safely mixed in with the normal patterns of life, requiring controlled spaces within which the product can be put through its paces without danger of contact with other systems and persons. A U. S. Air Force flight test base like Edwards is an example, and the auto companies maintain test tracks in cold and hot areas of the country. It sometimes happens that these spaces are insufficient. The auto companies certainly use the normal roadways for some forms of testing, but they, of course, abide by all pertinent laws in so doing. In testing some cruise missiles (sans weapon), it was necessary to overfly spaces beyond the confines of flight test bases in order to obtain sufficiently long legs and realistic terrain. From time to time, high-altitude remotely piloted reconnaissance aircraft have been flown in commercial airline corridors high above normal traffic flow, joining up two range spaces. In one of these cases, the aircraft had an engine failure, causing the remotely piloted aircraft to come down through the commercial corridor, luckily without incident.

Often the product will have to be instrumented to collect data that does not register on the human senses or because there is simply so much data that no reasonable number of people could observe and understand the meaning of it all in real time. This often involves instrumentation installed in the product under test, but also the spaces within which it is operated must often be configured for compatibility with this instrumented environment. This may involve only tracking capabilities but often instrumentation data reception and processing as well.

10.1.2.2.2 Operational Test and Evaluation

Operational test and evaluation (OT&E) is accomplished by the customer in accordance with mission needs that dictated the procurement of the system adjusted over the development life of the system for its evolving characteristics. In Chapter 1 the seed was planted about the potential for serious program problems concerning user requirements. On DoD programs, the user must develop a set of requirements, often in a document called an operational requirements document (ORD) or system requirements document (SRD) in the past, and more recently referred to as a joint capability document (JCD), initial capability document (ICD), or capability development document (CDD). This document may not state the requirements in a verifiable form, but rather as goals and thresholds. These requirements are subsequently crafted into a system specification in a verifiable form with all requirements appropriately quantified. In the process of creating the system specifications, the acquisition customer must make hard choices concerning affordability versus performance expectations of the user community. Where the two were in conflict in the past, performance would always win, but in the post-Cold War world, more often than not, affordability wins.

Today DoD lives in a world of cost as an independent variable, meaning that cost is traded along with performance. If capabilities are trimmed from user operational needs in producing the system specification on a DoD program, the user community may harbor ill will toward the acquisition agent and contractor that can outlast even a long development program. During OT&E, when the user community finally gets a chance to examine the system to "their requirements," the user may very well use the original user requirements documentation content as the basis for their operational evaluation, leading to a system failure. Now, the fact that the testing involved stresses and demands beyond those called for in the contract that yielded the system will in time be discovered, but the initial ill will generated toward the contractor and acquisition office may take some time to clear.

Unfortunately, the system engineering standard EIA 632 encourages this result in its identification of two kinds of validation. The first, requirements validation, evaluates the condition of risk in the requirements prior to the attempt to synthesize them in a design. The second, product validation, seeks to prove that the system satisfies users needs. The author refers to the latter as part of the system test and evaluation verification work at the system specification level, with DT&E focused on the engineering condition in accordance with the contract and the OT&E focused on mission capabilities that should also be constrained by the constraints of the contract that defined the system. It is very important early in the program during system specification development and what EIA 632 calls *requirements validation* (and the author calls simply *validation*) that the user requirements document content that will not be respected on the development contract be clearly identified, not just simply deleted from the system specification content. These requirements should be captured in a separate document or where their deletion results in a failure of traceability between user requirements document content and the system specification that the failure of trace be annotated by a comment that the user requirement was dropped for affordability concerns or a comment to the effect that the user requirement was converted into a desired goal rather than a requirement. Ideally, these cases would be consciously agreed to early by the user community, based on financial realities, and used as a basis for preplanned product improvement, block upgrades, and engineering change proposals as time and money permit.

The contractor would do well to maintain a list of these excluded user requirements and work toward finding ways to bring them under the contract through engineering change proposals or to satisfy cases of consideration that the customer has claimed for contractor benefits previously received under the contract. At every major review this list should be noted. The ideal way of capturing this list is to have loaded the user requirements document into the requirements database, and the traceability between that document content and the system specification will tell exactly what user requirements have not been included in the contract. These kinds of requirements are also consistent with the DoD adoption of the spiral development program management sequence model, where very risky requirements may be set aside for a later spiral pending availability of an enabling technology or funding.

Another difference between DT&E and OT&E is that OT&E is generally run on production equipment and software so instrumentation equipment is commonly not installed. One common result is that, if incidents occur resulting in loss of product or life, it will be more difficult to determine the ultimate cause than during DT&E, where full instrumentation is likely included.

10.1.2.2.3 Initial or Interim Operational Test and Evaluation
If the DT&E reveals a need to make significant changes to the system and these changes will take some time to develop and implement, the residual

assets from DT&E and any other end items produced to satisfy the customer's initial operating capability (IOC) requirement will not be in the correct configuration to use as the basis for the OT&E. This may lead the user to alter the original plan for OT&E to focus on those parts that are not affected by the changes early on and postpone test elements that are affected until the changes can be implemented. The resultant testing may be called *initial* or *interim operational testing and evaluation* (IOT&E). This can happen where the user is very motivated to take delivery of the capability to help deal with a near-term problem. Risks are taken in the development schedule by closing any schedule gaps as much as possible, and one of these gaps is commonly caused by a respect for the potential for faults that will be discovered during DT&E. In a user-driven schedule, risk concern is mitigated by close coupling of the D&E and OT&E; the product entities available for OT&E initially will not be in the correct configuration if serious problems were detected in DT&E. This same effect can occur between the end of qualification and the start of DT&E.

During the flight test (DT&E) of one low-altitude photo reconnaissance remotely piloted aircraft, the author can recall that it was shown that the aircraft could not be flown at high engine RPM on a low-altitude run because one of the two camera windows (one for a cross-track camera and one for an along-track camera), both of which were located on the underside of the nose camera compartment, interfered with the engine inlet air stream. When deployed, that aircraft had to fly at low altitude and lower than desired speed through the thickest antiaircraft capability that had ever been assembled at the time. Work started immediately on another version that only included a cross-track camera and lower drag wings to improve survivability.

10.1.2.2.4 Follow-On Operational Test and Evaluation

As the final resources begin to become available to the user subsequent to the DT&E changes discussed previously, the user may choose to transition to a final set of test planning data that corresponds to availability of fully operational assets, leading to identification of the test activity in the context of a follow-on activity (follow-on operational test and evaluation, FOT&E). Another way to get into this kind of test and evaluation program is that during OT&E user testing uncovers problems that did not surface during DT&E, because of the different perspectives of the people and organizations responsible for DT&E and OT&E. DT&E plans are structured based on an engineering perspective, whereas the OT&E is structured based on a user's mission perspective. The user will set up test sequences focused on the missions that are known to be needed, and these differences may expose the system to modes not evaluated during DT&E. These tests may uncover system problems that have to be fixed and follow-on operational testing accomplished after the fixes are installed.

The cost of making the changes driven by OT&E testing may or may not be properly assigned to the contractor. In order to allocate the cost fairly, it must be determined whether the testing attempted is appropriate for a system as described in the system specification. It is possible that the user may extend the mission desires beyond the mission capabilities they procured. This problem can easily materialize when a procurement office acquires the system and an operational unit operates it. The people in the latter commonly have very little interest in the contractual realities within which the procurement agent must function. They are strictly focused on their mission, and if the system will not satisfy some portion of it, the system should be changed. It may even be the case that the system in question provides better capabilities than required, but it still falls short of the capabilities that the user has extrapolated from the capabilities observed during DT&E that suggest other opportunities not yet satisfied nor covered in the contract. Users march to a different drummer than procurement people, and a contractor must remain alert to ensure that they are not steamrollered into making changes that are actually out of scope for the current contract.

10.1.3 Other Forms of System Testing

10.1.3.1 Quality and Reliability Monitoring

On a program involving the production of many articles over a long period of time, the customer may require that articles be selected from time to time to verify continuing production quality and reliability. This is especially valuable where the product cannot easily be operated to develop a history. A good example of this is a cruise missile. It would be possible to configure selected vehicles off the line for flight test purposes, involving adding command control and parachute recovery systems, but while these changes would permit the very costly vehicle to survive the test, they would invalidate the utility of the test. In order to fully test the missile we would have to cause it to impact a representative target and explode (conventional warhead assumed here, of course).

Military procurement of missiles for air-to-air, air-to-ground, ground-to-ground, and shipboard variations commonly involves purchase of blocks of missiles. Since the military cannot determine how many wars of what duration they will be required to fight, these missiles end up in storage bunkers and warehouses. The service may require that missiles be pulled from stock at random and actually flown to destruction from time to time. These actions, when the missiles work, encourage confidence in the store of hardware. These events also offer user units a great opportunity to conduct live fire training, so these two goals can be connected to great benefit, but with no small cost penalty.

10.1.3.2 System Compatibility Test

Ryan Aeronautical developed a complete remotely piloted reconnaissance system for the U.S. Navy in the late 1960s. This was a shipboard rocket-assist launched aircraft using an adapted ASROC booster installed on the vehicle, located in a near-zero length launcher angled at 15 degrees to horizontal. Ryan had been supplying the Strategic Air Command with this equipment for air launch from DC-130 aircraft for several years at that point, but had never completely outfitted a new user. Past operations grew into their then-current capability initially in the presence of a complete Ryan Aeronautical field team, with adjustments over time lost in the corporate knowledge. The program manager for the new program correctly sensed that there was a risk that the development team might not be capable of detecting from a purely intellectual perspective some subtle problems that might get into the design, so he required the performance of a system compatibility test. The author was asked to design and manage this test because he had experience in the field with the system and all of its support equipment.

The series of tests uncovered many small problems with vehicle and support equipment design, but did not disclose several very serious ones that were only exposed in the subsequent flight test. The principal value of the test was to confirm the soundness of the maintenance process as a prerequisite to entering flight test. Most of the problems that survived until flight test entailed operational aspects that could not be examined during the compatibility test.

Some things that were discovered:

a. The system was to be operated on an aircraft carrier at sea continuously for weeks at a time. When the vehicle returned to the ship after its mission, it was recovered through parachute descent to water impact, where it was picked up by the helicopter otherwise used for air-sea rescue on the carrier and placed on deck for postflight work that included fresh water decontamination of the aircraft and engine. An engine decontamination tank was provided to flush the external and internal parts of the engine with fresh water to remove the corrosive salt water residue. When an attempt was made to place the engine in the tank, it was found that it would not fit. The problem was traced back to a conversation between the lead support equipment designer and the new engineer given the responsibility of adapting a previous tank design to this engine. The original tank design had been made for a J69T29 engine common to the Ryan target drone. The lead designer told the new engineer to go down to the factory floor where there were a number of the correct engines awaiting installation in vehicles. The new engineer, not knowing that there were several models in production with four different engines happened upon a group of larger Continental engines used in a different model that did not have the same external interface and dimensions, and used this engine as the model for his design. Therefore, the adapter the new engineer created did not fit the Continental J69T41A engine used in this model. It had to be modified.

b. A number of other incompatibilities were uncovered, but the author has forgotten the details.

c. The test also acted as a technical data validation, and numerous problems were discovered in technical data and corrected.

d. In general, the series of tests confirmed that most of the equipment worked in a compatible fashion as a prerequisite to entering flight test.

Some things that were not discovered:

a. The Continental engine representative said that the J69T41A engine would not accelerate to launch thrust after start without an engine inlet scoop. The launcher design team located a retractable scoop on the front of the ground launcher such that it could be retracted out of the way subsequent to engine start and not be struck by the rocket motor moving past during launch. During the compatibility test (which included a launch simulation but not an actual launch), the scoop worked flawlessly while the vehicle sat in its launcher with the engine running at launch RPM. Several weeks later, the system went to flight test operated out of Point Mugu, California. On the first launch, the rocket motor blast tore the engine inlet scoop completely off the launcher and threw it some distance. It was quickly concluded that the inlet scoop was not required and the system worked flawlessly for many operational launches from the launching ship (USS Ranger) in that configuration during an operational evaluation.

b. A special circuit was designed for this vehicle such that if it lost control carrier for a certain period of time, it would restore to a go-home heading so as to regain a solid control carrier at shorter range from the launching ship's UHF line-of-sight control system. The only time this circuit was ever effective during operational use, it caused the loss of the vehicle rather than the salvage of it. The remote pilot during that mission was experiencing very poor carrier conditions and kept trying to issue commands to turn the vehicle back toward his control aircraft (an E2 modified with a drone control capability). What happened was that every time the carrier got into the vehicle momentarily, it reset the lost carrier go-home heading timer so that it never was triggered. Poor control carrier conditions caused the vehicle to only go into a turn during the very brief moments when the controller carrier got through. Since remote turns were not locking, the effect of these turn commands was almost nonexistent and the vehicle continued to fly further from the controlling aircraft. The controlling aircraft could not follow the vehicle to reduce the distance because the vehicle was overflying North Vietnam. Finally, the vehicle fueled out over Laos and crashed.

c. In preparation for launch, the jet engine was first started and brought up to launch RPM while other prelaunch checks were conducted. There was only a very brief time before too much fuel would be burned, resulting in an unsafe center of gravity for launch, so the launch procedure was stressful for the launch operator. Upon launch command with the jet engine running at launch RPM, the ASROC rocket motor was fired, shearing a holdback. The jet engine thrust vector was supposed to be balanced with the rocket engine thrust vector in the pitch axis so that the vehicle departed the launcher at the 15-degree angle at which it rested in the launcher. On the first launch at Point Mugu, California, the vehicle departed the launcher and roared straight up through an overcast and descended shortly thereafter behind the launcher in a flat spin, crashing fairly harmlessly between the launch control building and the launcher. System analysis had incorrectly estimated the pitch effect of jet engine exhaust impingement on the rocket unit slung under the vehicle in the exhaust stream of the jet engine and elevator effectiveness at slow speed, so the balance condition was beyond the control authority of the flight control system. It was unfortunate that a few weeks before this a Ryan target drone had been ground launched from this facility and the RATO unit, with an undetected cracked grain, had stutter fired, blowing up the drone just off the launch pad on its way to falling through the roof of a hangar on the flight line some distance away. But more problematical was the fact that the base commander was in the control building watching the launch of the new vehicle, and his wife and child were parked in the parking lot in a new Cadillac convertible between the control building and where the drone impacted. This probably made it a little more difficult to regain range safety approval for the next launch. A couple of years later, essentially the same crew launched a similar vehicle at a base in the Sinai desert from a depressed hangar access taxiway. The vehicle just barely cleared the ground level at the top of the ramp, because the control engineers had erred slightly in the opposite direction. At least in this case, the vehicle survived. On that launch very likely General Moshe Dayan, who witnessed the launch, had some trouble figuring out why these Ryan engineers were laughing so loudly as the drone missed the ground. He had not heard the control system engineer say, "I guess we overcorrected a little." Testing is necessary because it is simply not possible to analytically determine all outcomes.

d. The rocket igniter circuit was supposed to have been completely isolated from the rest of the electrical system. It tested okay during the compatibility and electromagnetic interference (EMI) tests. On the last launch of the flight test program from the USS Bennington, the igniter was set off when the launch control operator switched on the fuel boost pump in preparation for jet engine start. The rocket fired without the jet engine operating, and the vector sum drove the vehicle into the ocean just forward of the launcher. Subsequent EMI testing revealed a sneak circuit, which had not shown up in the previous launches and tests and which had to be corrected before operational use.

e. In loading a vehicle onto the launcher, one had to hoist the vehicle from a trailer in a horizontal attitude with the rocket attached and then rotate it to 15 degrees nose-up to place it on the launcher rails. A special beam was designed with a traveling pickup operated by an air-driven wrench. This worked well during the compatibility test. On one launch on the USS Ranger, however, the user experienced a failure of the rocket motor to fire during an operational launch. This is one of the worst things that can happen with a solid rocket, since you don't know if it may ignite subsequently. The rocket could not be removed while the vehicle was in the launcher, so the whole apparatus had to be moved into the hangar deck, where many aircraft parked loaded with fuel had to be moved, and the aircraft hoisted from the launcher with rocket attached with the hoist beam pickup point set for a 15-degree angle. The beam was supposed to permit movement of the pickup point under load to a horizontal position, but in this case it did not move. With the pickup point jammed too far forward, the vehicle had to be manhandled into a trailer, where the misfired rocket could be removed and pushed overboard.

10.1.3.3 Certification

Many product fields require a certification process endorsed by some government agency. Commercial aircraft require FAA certification. Some medical apparatus requires certification by the Food and Drug Administration. These certification tasks can generally be mixed with the normal system test and evaluation process so that many tasks need not be repeated. In these cases you have to produce evidence of compliance with technical product requirements as well as produce evidence of compliance with certification criteria. Generally, the focus is on public safety in these trials. There may be a statistical element in this process requiring a statistically significant sample that can extend the cost and time required.

The certification process also extends to people performing some professions. While it is possible to acquire employment as a system engineer in industry without an International Council on Systems Engineering (INCOSE) certification, it does help. This status is achieved by submitting an application revealing experience and by passing a test where the questions are derived from the INCOSE System Engineering Handbook. Many professions involving a position of trust require that the credentials of a person applying for that status be verified by a respected agency as achieving a well-defined level believed to reflect that the person can be trusted to perform an activity well.

10.1.4 The Grand Plan

The first verification document most often required is variously referred to as a test and evaluation master plan (TEMP), integrated test plan (ITP), system test and evaluation plan (STEP), or a half dozen other names (see Table 9). By whatever name, it is intended as an overall management plan with which to control the system testing work. It sets policy and identifies program goals for testing. The TEMP identifies resources needed in terms of special test articles, support equipment, facilities and ranges, and instrumentation. It does not fit in with the same kind of plans discussed in previous chapters to define the specific objectives for particular verification actions, but that detailed planning based on the specification content is required and we will discuss it shortly.

10.1.5 The Limits of Intellectual Thought

None of us human beings like to think that our intellectual capabilities are suspect, but that is the case. There are limits beyond which pure thought

Table 9 Test and Evaluation Master Plan Outline

PARA	TITLE
	Executive Summary
1.	System Introduction
1.1	Mission Description
1.2	System Description
1.3	System Threat Assessment
1.4	Measures of Effectiveness and Suitability
1.5	Critical Technical Parameters
1.6	Environmental Safety and Occupational Health
2	Integrated Test Program Summary
2.1	Integrated Test Program
2.2	Management
3	Development Test and Evaluation Outline
3.1	Development Test and Evaluation (DT&E)
3.2	DT&E to Date
3.3	Future DT&E
4	Operational Test and Evaluation Outline
4.1	Operational Test and Evaluation (OT&E)
4.2	Critical Operational Issues
4.3	Operational Impact Assessment (OIA)
4.4	Special Considerations
4.5	Operational Test and Evaluation to Date
4.5.1	Post OA II Events, Scope of Testing, Scenarios and Limitations
4.6	Post OA II Evaluation (POA)
4.7	Live Fire Test and Evaluation
5	Test and Evaluation Resources
5.1	Test and Evaluation Resources
5.1.1	Test Articles
5.1.2	Test Sites and Instrumentation
5.1.3	Test Support Equipment
5.1.4	Threat Representation
5.1.5	Test Targets
5.1.6	Operational Force Test Support
5.1.7	Simulations, Models, and Test Beds
5.1.8	Special Requirements
5.1.9	Test and Evaluation Funding Requirements
5.2	Projected Key Test and Evaluation Resources
A	Annex A, Bibliography
B	Annex B, Acronyms
C	Annex C, Points of Contact

cannot be relied upon, and the limit is related to situational complexity. Airborne separation of stores from aircraft is one case of complexity that is better proven through test after the best analysis available is applied.

A former U.S. Marine fighter pilot once told the author about an experience on a bombing run during the Korean War; after he released the bomb from an F4U Corsair, he noticed the bomb rolling out the bottom surface of the wing around the wing tip and back up toward him on the top of the wing. He wasn't sure how he had finally gotten the bomb to separate but thought it had something to do with how loud he was screaming. Weapons release from this aircraft had been thoroughly studied and subjected to flight testing, but apparently every flight envelope, loading, and attitude combination had not been exercised for that particular weapon.

Despite careful analysis and wind tunnel work, there were several Ryan Aeronautical remotely piloted reconnaissance aircraft air launch mishaps from DC-130 aircraft during the 1960s. A Ryan model 147 J low-altitude photo reconnaissance aircraft took off the number 3 engine of the launch aircraft due to a flight control system problem. Another came apart over the top of the launch aircraft. It happened that the same USAF major was the pilot on these two flights, leading him to inform the author (with a little irritation in his voice) while they were sitting out an operational mission hold on the tarmac at Da Nang air base in South Vietnam, "You know, Grady, I was the pilot of the DC130 that lost an engine to one of your birds, and I was the pilot of the DC130 that one of your birds blew up over, and I am going to be the pilot of a DC130 when one of your @%# birds kills me." The author is happy to report that the major survived the combat mission that day, especially because the author was on it, and the pilot even retired alive much later. A Ryan model 147H high-altitude photo reconnaissance bird initially refused to depart the launch aircraft when commanded to do so because it had so much lift. Spoilers had to be added to encourage separation at launch and to cause the aircraft to descend from high altitude for recovery at mission end.

In order to make absolutely certain of proper separation, the USAF/General Dynamics Convair Advanced Cruise Missile program called for the launch and jettison of dummy vehicles (reinforced concrete shapes reflecting actual vehicle mass, center of gravity, and surface shape and smoothness) from all launch stations on B-1, B-2, and B-52 aircraft prior to starting flight test with the actual flight test vehicle.

Combinations of environmental stresses are another complex situation that is often hard to imagine through pure thought without actual trials.

The USN/Librascope MK111 ASROC ASW Fire Control System design included a serial digital differential analyzer produced at a time when rotating memory drums were popular. The unit worked very well in the plant and aboard the first installation in a destroyer. The ship had one more trial to undergo on the east coast before it sailed for the west coast to enter ASROC missile DT&E testing. During depth-charge testing of the ship, the rotating memory drum sheared off all of its read-write heads. The problem was fixed and the system provided good service, but the next order of ASW systems for destroyers replaced the MK111 with the MK113, which solved the ASW problem using an electronic analog computer.

Human response to unusual stimulus is an uncertainty at best, despite elaborate efforts to prescribe normal and emergency procedures. Our best efforts to analyze these situations are often based on the humans acting appropriately and precisely in accordance with the procedures. Actual trials under safe conditions (which may be an impossible combination) may reveal new insights about human responses as they are played out on a particular system.

10.1.6 The Details

The TEMP can be supplemented with the detailed documentation suggested in prior chapters on item qualification, driven by the content of the item performance specifications and acceptance testing driven by the content of the item detail specifications. The primary concern of the TEMP, however, is the system testing. This planning process works exactly as it is described in Chapter 8 for qualification. System specification requirements ideally would have verification requirements written for them concurrently and a method defined for verification. All of these requirements would be swept into the verification compliance matrix, where engineers responsible for designing the verification plans will allocate line items or strings to specific system test and evaluation tasks. Each unique task is assigned to a specific engineer, who will have the responsibility to collect all of the requirements assigned to his or her task and fashion a test and evaluation plan and procedure that will yield evidence of compliance of the design with the specification content.

Note that this process does partition all test and evaluation work into individual tasks and that any time we partition a whole into parts we must integrate and optimize the resultant plan to ensure that we have arrived at a satisfactory whole.

10.2 SYSTEM TEST AND EVALUATION IMPLEMENTATION AND MANAGEMENT

10.2.1 System Test and Evaluation in the Beginning

While this chapter includes the word *test* in the title, the chapter extends the words *test and evaluation* to include the other verification methods as well. All five methods are commonly appropriate at the system level.

Like all verification work, system test and evaluation starts with a good specification, in this case the system specification. The management process must therefore begin with the development of this specification. Early management aspects are very simple but necessary, such as ensuring that the system specification includes a verification traceability matrix, that it is completed, and that it includes a mapping to the verification process requirements, ideally located in Section 4 of the system specification. From this point forward, the management aspects must focus on encouraging the timely development of a compatible instrumentation plan and design and development of a minimum number of effective verification tasks characterized by affordability and timely completion.

10.2.2 Instrumentation Management

Item verification seldom requires special instrumentation contained within the design of the product, but it is very common in end items subjected to system test. In some engineering organizations, the instrumentation engineering is started as the product engineering tapers off, meaning that the instrumentation engineers must work with whatever design was previously concocted. Often this is a severe limitation on the instrumentation engineers. Ideally, an instrumentation engineer would be accepted as an early team member on what the author calls the program integration team, to begin the instrumentation requirements development and design planning.

Generally, the instrumentation parameters should be decided by the engineers from the several participating domains, based on their concerns for their design and their ability to satisfy system and end item requirements. They should include parameters that will collect data needed to make decisions about the adequacy of their design. It is unwise to limit interest to domain engineer inputs, however, as these will all be domain oriented. There will be some valid system parameters that the specialists will not identify. Thus, the evolving parameter list should be scrutinized by someone with system knowledge and interests for the purpose of integration and identification of system-level parameters. If the program is organized using cross-functional teams including a system team, called a program integration team (PIT) in this book, the teams should participate in this parameter list development, with the PIT taking the parameter integration role. The parameter list is, in effect, the instrumentation parameter requirements document and should include the range of values and corresponding units required.

Program management must recognize the need to integrate the instrumentation design process with the normal operational product design such that optimum design solutions are discovered early and implemented. In some cases this may entail permanently imbedding instrumentation designs in the operational product, as these designs may enable test equipment connections at the module, component, and subsystem levels. In other cases, it may simply be less costly to continue including a circuit introduced for instrumentation purposes in regular production units even though it has no useful purpose in normal operation and maintenance. Another potentially useful aspect of maintaining instrumentation design in a production article is that the system may at some later date have to enter reverification and this capability can support that testing.

Instrumentation systems commonly consist of sensing connections or special instrumentation sensors, a means to concentrate the data for efficient transmission, and some way to communicate, observe, and record the collected data. Because the product is commonly some form of mobile entity such as a wheeled vehicle or aircraft, the data must often be transmitted by some means from the article under test to a remote instrumentation station where the data is processed and recorded for subsequent analysis. A fairly common arrangement today entails many sensory data sources that are combined into a single signal by a multiplexer, which acts as the modulation signal for a radio transmitter. The radio signal is radiated from a unit under test from one or more antennas. If the test article maneuvers relative to the receiving station, antenna switching may be necessary to ensure that no matter the attitude and orientation, a solid instrumentation signal will be received by the station. The station, of course, must receive, demodulate, demultiplex, and record the instrumentation data faithfully.

Where the testing entails a potential need to intervene into test operations based on instrumentation observations, real-time instrumentation readouts may be necessary and be monitored by qualified engineers capable of detecting and diagnosing observed problems and communicating with those who have the capability of intervening. This is very common on space transport systems, especially those involving crewed flight, where even the astronauts are instrumented for medical observation. Remotely piloted systems also require, during system test and evaluation, a means to control the article under test for safety destruct purposes if nothing else. Most often, a remotely piloted system will require some form of command control during system test and operational use as well. The command structure may be slightly different between test and operations, but the testing must also verify that the command control system performs as specified, so one should not encourage a lot of differences here.

10.2.3 Reaching a State of DT&E Readiness

Development test and evaluation is conducted by the contractor to show the degree to which the produced system satisfies the system requirements agreed upon in a contract between the contractor and the acquisition agent of the customer. Prior to its opening act, all of the components that comprise the system should have been qualified in some way for the application. Every task in the process should have been planned and procedures developed, and the linkage established between plan content and the requirements in the system specification, so it can be shown that the DT&E process will comprehensively evaluate the system relative to that standard. Many of the tasks in the DT&E will involve very costly entities and sometimes offer considerable risk to life and limb as well as loss of the costly entities. Thus, there is a need for proceeding with caution in implementing the process with due regard for system safety. It may be necessary to implement an effective test readiness review for each task.

If the system involves a flight article, it will require a special test range on which to accomplish the DT&E. Examples of these ranges include Edwards Air Force Base in California and the Navy's Patuxent River facility in Maryland. The enterprises that develop systems commonly do not own such facilities, with the space, instrumentation, and supporting facilities necessary to support DT&E. An exception would be the auto industry, where companies commonly have access to their own test tracks. DT&E may, therefore, have to take place on a range facility operated by some element of the customer. Access to these ranges is difficult, so long-range planning is necessary. These facilities are in great demand, so once DT&E has begun it will have to proceed from start to planned finish with no significant delays while problems are studied, understood, and fixed. DT&E appears in the life-cycle process as shown by the heavy bordered block in Figure 71.

Leading up to this point, the TEMP was prepared early in the program and updated during design to provide policy and management guidance in

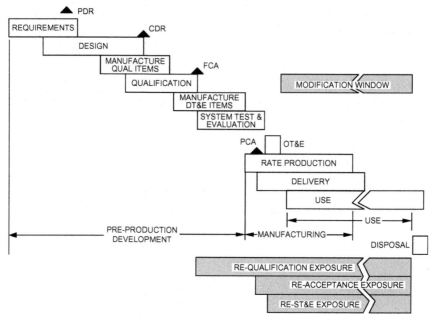

Figure 71 Overall system test and evaluation situation.

the implementation of the DT&E work. Also, the system specification content has been used as the basis for detailed DT&E task planning. As in qualification and acceptance, the DT&E planning process has entailed mapping sets of specification content (requirements in the system specification in this case) to DT&E tasks, identifying resources needed for each task, and providing a "how-to" procedure to follow for each task. Each DT&E task should be placed on a flow diagram coordinated with a schedule, or better, in a project planning tool that can handle process flow as well as resource management using PERT or CPM principles.

There is a great temptation to plan the system test and evaluation tasks based on the product design, but this is a serious mistake. The intent is to produce evidence on the comparison of the product with the requirements the product design is intended to satisfy.

10.2.4 A Normal Evolution

The key system test and evaluation management document is the verification task matrix component dealing with system test and evaluation (all other tasks suppressed). As in item qualification and item acceptance, this matrix lists every task in the system verification process, identifies the principal engineer, identifies and provides status of the task plan, procedure, and report, tells the budget available for the task, and gives task schedule information. The system test and evaluation planning and procedural data for each task provide a list of the resources needed and the required detailed steps for each task. The report for each task should provide a clearly stated record of results from which it can be determined if the system design complied with the requirements the task was inspecting the system for. The task matrix should be maintained as a living document, ideally on a computer rather than released periodically in paper. It will provide management with a list of all of the system test and evaluation tasks and their status.

Much of the verification work accomplished at the system level is inflexibly sequential because it is necessary to progressively organize the verification work. Before we dare take off in a new airplane, it is a good idea to prove that it is possible to taxi at high speed on the ground. Before we prove that the new aircraft can maneuver in high-g turns, it is a good idea to prove that it will fly in a stable flight regime. So, a system test and evaluation program falls into place as a progressively more stressful series of tasks, each answering vital questions while building confidence for subsequent tasks.

All of the system test and evaluation tasks are not accomplished by the test method, and these others require some management interest as well. Some examples of system-level analyses are the following:

a. The reliability, availability, and maintainability prediction is accomplished by engineers determining the appropriate figures for elements of the system based on their design and determining system values from their mathematical models into which the individual figures are introduced.

b. The mass properties engineers determine end item weight and center of gravity through calculations based on actual designs and by taking measurements of actual hardware.

c. An economic analysis may be accomplished to determine life-cycle cost.

d. Network analysis may have to supplement testing of networked systems under less than full capacity conditions.

In all of these examples, the figures determined based on the actual design are compared with the allocations made prior to detail design, to determine if the designers met their target values. If some designs are better than needed, it may be possible to reallocate values to avoid redesign of other elements that did not quite make their allocations. When a design fails to meet its target values, management should consider allocating any available margin, changing the requirement, or redesign actions that would cause compliance.

Note that the analytical verification evidence will most often be available long before that derived from test, demonstration, and examination, because analytical tasks require no hardware build time and can be concluded as soon as the design is set.

At least some of the system test tasks will be sufficiently critical to program success and possibly entail serious problems if they have to be repeated, so you should hold test readiness reviews before initiating work. In this review, we should make certain all of the personnel slated to witness the test are present, that the resources are all in place and working, and that the article under test is in the proper configuration. Verify that any supporting resources such as range equipment and space are in readiness in accordance with the plan and schedule. Management should require those responsible for accomplishing the task to walk through the paces and demonstrate any time-critical parts of the task. This review is similar to what a movie director would feel necessary before a big action scene where the set used throughout the movie is burned down. If it doesn't go perfectly, they may have to build a whole new set. The same thing can happen in program-critical testing where costly resources are going to be consumed.

The person responsible for managing some part of the DT&E must ensure that the planned tasks are implemented on schedule and that they stay on track on their way to completion. As each task is completed, it must be reported upon, and that report collected for the final DT&E report. There is no single task in the verification process that will yield more positive results than a program following the rule that all task principal engineers include in their report a copy of the specific list of requirements from the specification that the task must produce evidence of compliance for. Management should review reports, to ensure that no task report is approved unless it includes this information and that the evidence cited is clearly described and the connecting logic flawlessly communicated.

DoD has never identified a major formal audit for completion of DT&E paralleling the use of the functional configuration audit (FCA) at the end of qualification and physical configuration audit (PCA) at completion of the first article inspection. Some form of audit or meeting is encouraged between the acquisition agent and contractor, however, to reach a formal agreement on the degree to which the content of the system specification was proven to have been satisfied in the design of the system. When everything goes according to plan, it is but a happy management task to keep things on track to a conclusion.

10.2.5 Flexibility in the Face of Obstacles

For someone who has experience in DT&E, it is hard to imagine everything going according to plan throughout, no matter the quality of the plan. We are dealing with extremely complex issues in DT&E that are simply very difficult to intellectualize. When a discontinuity does occur in the planned work, we have to come to an understanding of the cause quickly and resolve on a path that will allow us to complete the tasks involved as close to planned schedule as possible. One approach to adversity is to adjust the relative schedule for some of the work to move problematical tasks to a later time when we believe a related problem will have been cleared, and move into the void later planned tasks that are not influenced by the observed problem.

Task rescheduling can be a problem, of course. Ideally, we would have set up the original sequence to encourage movement from the simple to the complex, safe toward hazardous, and easy to hard. If we find that we cannot accomplish some of the relatively simple, safe, and easy tasks, but could fill the space with some less safe tasks, we may be able to close up the schedule, but only by taking some risks we would rather avoid. For example, during DT&E continuing structural dynamics testing and analysis, it has been found that the wings of our aircraft cannot withstand the intended full loads that occur when the aircraft is flown through a turning climb pulling 7G. This does not mean that we have to delay the whole DT&E program. Pending an evaluation by competent structural engineers, we may be able to proceed with all planned initial flight testing focused on stable flight at various points in the airspeed-altitude envelope, interspersed where necessary with moderate turns and altitude changes. While this is going on, we will have to corroborate the problem and, if true, find a way to overcome the problem so as to permit entry into maneuvering flight tests as close to schedule as possible.

There are many other similar possibilities:

(1) An aircraft with external fuel stores capability may have a flawed design in the wing-mounted quick disconnect port such that it leaks when an external store is attached. We should be able to complete all slick wing flights while delaying entry into fuel pod flights until after the design is changed and the aircraft retrofitted. If the disconnect leaks even with no fuel pod attached, we can cap off the line internally if necessary.

(2) In early flights we discover that undesirable flight conditions exist at speeds much higher or lower than predicted. For example, we discover that stall occurs at a speed 100 knots higher than predicted at a particular altitude. There are two problems here that should concern us. First, we

can placard the flight charts with a higher stall speed and continue flight while we determine how to open up the envelope to the originally planned range. Second, we should inquire how our engineering process went astray to cause such a difference between predicted and actual performance. In system test we have a tremendous opportunity to observe the integrity of our engineering process. We would always like the difference between predicted and actual performance to be small, but to the extent that it is not, we should try to discover why it was not and possibly uncover a potentially serious process flaw that can be fixed, leading to more effective development efforts in the future.

(3) In the flight test of a new low-altitude photo reconnaissance camera, it is found that it cannot adjust rapidly enough to changes in the velocity/height ratio (V/H). It may be possible to reshuffle the flights to make early flights over fairly flat terrain while the camera company engineers work on the V/H problem. In another situation, a night photo reconnaissance aircraft development discovers that the night camera system has serious flaws, but the flight test of the aircraft based on pure aerodynamics can be completed in daylight conditions while the camera is perfected.

10.2.6 DT&E Reporting

DT&E reports can be prepared in the same pattern as previously described for qualification and acceptance in the form of individual reports for each task. Earlier versions of this book encouraged use of an integrated report capturing all reports within one set of covers, but the author finally concluded that this was a case of overengineering a simple problem. As previously noted, the reporting data should actually be captured in the same database system as all of the specification and verification documentation, with longitudinal traceability data included all the way from the requirements in Section 3 of the system specification in this case through the DT&E plans and procedures documents to the DT&E report paragraphs.

In DT&E there is another form of reporting that is often required by management of the development contractor, called a quick-look report by some. Clearly the DT&E tests are very important to management and the whole company. The program is in its final stages, nearing the time when the user will start taking delivery and a time when rate production will gear up, producing a steady profit flow to the company after many years of uncertainty. Management will often hang on every word from the DT&E site and will always want the latest information about what is going on. This can become a problem from the DT&E management perspective at the DT&E site. These are often very action-oriented people who want to get the job done, and they often do not well tolerate what they see as busy work. A good way to satisfy management desires for status with DT&E attitudes is to provide for a quick-look report that provides, as soon as possible after a test task, what happened and what it means. This could be done with a simple email today, giving the task ID and a text description of the task outcome.

If a task was considered a failure, and especially if the test article is no longer useable, as in a crash, it will require more than a simple email to report the results. Very likely some form of incident investigation will be required, not only by management but by the range. The range will insist on understanding the cause and what will be done to preclude such an outcome in the future. These kinds of incidents can cause long delays in the DT&E program and are examples of the situations where the program will have to be alert to ways to continue in a flexible fashion while solving a problem.

10.2.7 Logistics Support in DT&E

One of the problems that make it very difficult to maintain DT&E schedule is the poor logistics support available at the time it is conducted. There are few if any spares. Therefore, spares will commonly have to come from cannibalization of items from other end items in DT&E. If the plant is located fairly close to the DT&E site and the company has its own airplane, it can be used to fly the failed parts back to the plant for expedited repair, treating the plant like a depot. Alternatively, FedEx can be used or DT&E workers can carry the failed units back to the plant on commercial flights.

10.2.8 OT&E

Upon completion of DT&E and, ideally, completion of design changes encouraged by DT&E, production can begin on the high-rate production units that will be shipped to the user to begin their march toward an initial operating capability (IOC). As part of the process of approaching IOC, the user may conduct an OT&E. If the user need is sufficiently urgent, the acquisition agent may have authorized contractor entry into rate-production before the end of DT&E, recognizing that considerable cost risk could be realized if serious problems are uncovered in late DT&E. Clearly, this is a case of trading schedule risk for a potential cost risk.

The contractor seldom has a clear connection with the detailed results of the OT&E. The contractor will probably have to rely upon their acquisition agent passing on messages from the user regarding progress. The author can recall one case where the contractor could stay on top of OT&E proceedings. Teledyne Ryan Aeronautical had field engineers serving at all of the user sites that were involved in remotely piloted aircraft OT&E activity. Often the field engineers flew on launch missions, attended debriefings, and were involved in postflight maintenance actions, so they were able to keep contractor management informed about events of interest. These field engineers often had to be very careful in divulging the truth about OT&E events, just as they had to be when reporting from operation sites at other times. If a user unit crashed an OT&E vehicle, it was sometimes a great temptation for the field engineers to blame a bad design rather than a maintenance preparation step, remote pilot action, or launch control officer failing. In these cases, the field engineer would know the truth, but if they reported it in a way that could get back to the unit, that field engineer might find it very difficult to be effective in the job from that point on. In such cases, the field engineer can beat the best path by encouraging the user to include only truthful statements in their status reports.

10.2.9 Test Results Applications

The results of system testing are first useful in verifying that the system complies with system requirements. In addition, the results of system testing are useful in confirming the validity of models used to create the system. Throughout the development process, while the system was only an intellectual entity defined in a growing list of engineering drawings, models and simulations, there was no physical reality. These models and simulations were built based on some assumptions as well as rational assessments based on science and engineering. We should have done our best to validate these models and simulations as early in their life as possible, but the availability of the complete system is the first time that some of the features of these objects could be fully validated.

If system testing confirms that the models and simulations predicted performance that is actually observed in operation of the system, then we have a very valuable resource on our hands and every right to be very proud of our development effort, or dumb luck. The reader may conclude that since the system is already built, it is of little consequence whether or not the models agree with its operation so long as we are satisfied with system operation.

Why don't we simply get rid of all of the other representations of the system? We are finished.

The reality is that this system will have a long life, and in that long life it will experience phenomena that we may not understand immediately. If we have available to us a simulation that has been validated through system testing, then we can manipulate the simulation to help us understand the cause for the problem observed in a very controllable situation. Without these resources, we may be forced to conduct dangerous flight tests to observe performance aberrations. Also, the simulation will permit us to examine a larger number of a wider range of situations in a much shorter period of time than is possible through actual operation of the system.

But, more importantly, if the actual test results are significantly different than our various representations predicted, then there is something wrong with the development process we applied. Ideally, test results would reflect precisely what our analyses predicted. This result should give us great confidence in the stream of logic supporting the design.

So, we should make a conscious effort to correlate the results of system testing with any models and simulations we may need for future use. In some cases they will not agree, and the cause could be either the simulation is incorrect or we could have interpreted the results of system testing incorrectly. In any cases where there is not good agreement, we should look for a cause on both sides of the divide and make adjustments in our development practices and machinery to the degree possible.

10.2.10 System Test and Evaluation Closure

System test and evaluation should cease, of course, when it is complete as planned or formally modified, provided it has achieved the data needed to make a sound technical decision about the readiness of the system to be deployed in achievement of the customer's original or contractually adjusted needs. Like all program tasks, even those as much fun as system test and evaluation, it must be brought to closure as early as possible because it costs a lot of money. There is no formal major review commonly recognized for system test and evaluation closure as there is for item qualification in the form of the FCA. There should be one held, however, under whatever name you may choose.

This review should be structured like the FCA outlined in Chapter 8. After an introduction providing an overview of the whole test and evaluation process, each task, or some selected subset of them, should be briefed with an identification of the requirements to be addressed, the actions accomplished to establish the proof, and an interpretation of the results. Any cases of failure to perform to required values should have been investigated and a fix implemented, and any of these actions should also be reported upon.

In the case of a formal certification process, as in FAA flight certification or FDA certification of a drug, there are formal and regulatory wickets through which one must pass to formally bring about the desired result, certification, which means that the product has been found to be effective relative to its claims or safe for public use or consumption.

10.3 ITEM AND SYSTEM REVERIFICATION

10.3.1 Reverification Exposed

Reverification is the act of verifying that a product item satisfies its performance (development or Part I) or detail (product or Part II) specification content or that a whole system satisfies the content of the system specification subsequent to it having been attempted or accomplished at some time in the past. This chapter will focus on the item qualification aspect of this process, but it applies equally well for acceptance and system test and evaluation; those areas will be summarized after a detailed treatment of the item requalification process.

Chapters 8 and 9 as well as the preceding portion of this chapter were intended to deal with the first time a new product passes through the item and system verification processes. Are there cases where a product must pass back through the verification process? There are several sound reasons why this might have to occur, at least in part. This chapter systematically identifies those cases, and for each case it gives a set of response strategies involving specific steps for each strategy to accomplish the reverification action. This section on reverification was motivated by questions from a system engineer named Christine Rusch, who brightened up the room with her questions and mature insights into this difficult work during a verification tutorial given by the author for INCOSE in Las Vegas, Nevada.

10.3.2 Project Phase Subsets

With a couple of exceptions, reverification is a moot point prior to the initial qualification activity, so we are concerned with the program period subsequent to initial qualification, which can be divided into preproduction development, manufacturing, and product use. But the problem is not quite as simple as this, because all of these periods can run together in a program with even a moderately extended manufacturing time span or run because of manufacturing and use overlap.

10.3.2.1 Preproduction Development

The most straightforward rationale for qualification reverification is failure to pass qualification on first opportunity. Failure to pass qualification means that the design does not measure up to the content of the performance specification. The question then remains how to reach closure. There are several alternatives:

a. Change the requirements such that the verification evidence satisfies those requirements. This avoids any reverification action. It will be necessary to change the whole string of documentation from the specification through the related verification plans and procedures to the verification task report.

b. Change the design such that it will pass the current performance specification content. The engineering drawings or computer software code will have to be changed, but the rest of the data should be acceptable as it is cast. Also, the qualification process will have to be rerun at least partially, to prove that this new design does, in fact, satisfy the requirements subsequent to redesign.

c. Seek a waiver permitting the current noncompliant design to be sold for a specific period of time or number of articles while the design is upgraded to fully satisfy the original requirements. In the case of a commercial product with a large number of individual buyers, the developing company need not gain customer approval for this, but rather they simply reach a business decision on the effect on product viability if it fails to satisfy at least some of the previously agreed-upon requirements. In the case of a program with a single or very few customers, customer approval will likely be required. Department of Defense has very specific procedures for waivers. Also, the qualification process will have to be rerun at least partially, to prove that this new design does, in fact, satisfy the requirements subsequent to redesign.

d. During the development work, the customer often becomes aware of capability opportunities not previously envisioned and not included in the contract. Many people call this requirements *creep*, where the customer makes a series of small scope changes in the required capabilities of the product with or without formal contract changes. The contractor should protect itself from this potential cost hemorrhage by as early as possible gaining formal approval of the top-level specifications, rigorous protection of that baseline, and insistence on engineering change proposals as a way of upgrading system capabilities coordinated with fair payment for the impact.

e. Some developers evolve a final product through a poly prototyping process entailing two (alpha and beta), three (delta), or even four (gamma) versions of the final product, where the last in the series is intended to be the first production article. Each prototype should pass through some form of qualification process; thus each subsequent qualification is essentially a reverification. Generally, the capabilities of each prototype will escalate toward a final goal realized in the final prototype. The performance specifications for this kind of product may well include tabular requirements values with targets for each prototype. This is similar to a multiple spiral process in which the developer affects some form of verification on each spiral. The previous four rationales for re-verification are all reactions to unfolding circumstances. The poly prototyping development model permits crafting a predetermined plan effective across several product entities. This case may strike some readers as a perfectly normal agile development scenario, where the development team progressively closes on what the customer was looking for but was unable to perfectly describe initially.

10.3.2.2 Manufacturing

Once a product is in production, a new set of constraining conditions come into play. The fundamental difference is that some product has been produced; the program has passed through a barrier from pure intellectual achievement to reality. Physical product now exists, which adds a whole new representation that must be configuration controlled along with the engineering drawings and other representations such as simulations and models. If it becomes necessary to reverify for any reason, it is likely that it will extend through item qualification and acceptance as well as possibly system test and evaluation.

The only time during the manufacturing process where the capabilities of the product will come into question will be during item acceptance. Given that the product has passed qualification but fails acceptance, the fault may very well be in the manufacturing process. The qualification articles were very likely manufactured in an engineering laboratory by engineers and very experienced technicians and mechanics. The engineering drawings, process specifications, and manufacturing planning may not cover certain critical steps that the highly qualified people building and testing the qualification units were aware of.

One of the critical decisions to derive in this situation is whether the design is satisfactory – that is, to locate the fault in the engineering or manufacturing camp. Naturally, the pride and defensiveness of well-meaning people in each camp may hinder this decision-making process, but it is necessary to reach a clear understanding of the cause. We need the information that will permit us to decide the best course of action that is also affordable. Ideally, we could correct the problem and continue the manufacturing process without changing the engineering and requalifying the item.

There is a very important point connected with this problem. A failure in first article acceptance after a successful qualification brings into question the validity of the qualification evidence. It should have been a part of the functional configuration audit to make the case that the manufacturing of the qualification units is reflective of the planned-rate production process and that evidence accumulated from those units will be representative of production units. If it can be shown that the acceptance problem is rooted in an engineering problem covered up by a qualification unit features, consciously or innocently included, then the design must be changed or another option listed in the prior paragraph elected.

So long as a production run is in progress, its physical configuration audit should be evidence of process integrity. If production is interrupted or altered in some significant way for any reason for a sufficient time span, that integrity properly comes under question. There is a lot of gray here, but clearly if the production process is interrupted it should be subjected to a new first article process and closing physical configuration audit. This becomes a more emphatic need as the several factors become part of the changed situation: different practices, different people (workers and/or management), different facilities, different tooling, different customer, and changed product.

10.3.2.3 Product Use

A product in use may expose factual evidence that it does not in fact perform as required despite the results of the qualification process. In a Department of Defense procurement, the user defines a need, and an acquisition agent goes out and contracts for a supplier to develop and produce it. In the process of doing so, the acquisition agent often will discover that the difficulty of satisfying the need and amount of money and time available are not compatible. When the user finally takes delivery, they may still be thinking in terms of their original need and discover very quickly that the new system does not measure up. Contractual considerations like unavailability of sufficient funding tend to carry little weight with the user at this time.

As noted elsewhere in this book and the author's companion book *System Requirements Analysis*, both published by Elsevier, the developer should clearly list all user needs not covered under the contract and work energetically throughout the development period to close any gap between user needs and contractual coverage. At every opportunity involving the user during development work, the contractor should make any shortfall clear so that the user has no foundation for surprises as it begins operations, sometimes years later. Efforts by the user to characterize the system as failing to satisfy their needs should be based on factual information voiced within context with the contract through which the system was acquired.

There may be real areas of specification noncompliance exposed through user operation of the system that frankly were not exposed in contractor testing. The user will commonly take a very different attitude in testing the system than the contractor. The contractor will take an engineering view of their product, while the user will view the product from an operational mission orientation. Such problems can filter down into item design issues, stimulating reverification from an item qualification and acceptance perspective.

10.3.3 Temporal Considerations

Figure 72 shows the whole systems life cycle, and reverification will almost always occur in grand systems employment based on an appreciation for the system features relative to possible changes in the way it must be employed to provide best value. This is often motivated by the ways the competition (commercial or military) changes in response to introduction of the system. A need for reverification will almost always be necessary if the system is changed, of course, but an existing design may have to be reverified if it is determined that it can be used in a significantly different fashion. It is true, of course, that the earlier one discovers a need for reverification, the less costly it will be. The later one is exposed to a need for reverification, the more information that will have to be changed, and the greater number of decisions implemented that will have to be countermanded.

It is true that during the system development period, when the system is under development and prior to delivery of the system to the customer verification work accomplished can be found to be unacceptable, incomplete, or in need of change to coordinate with design changed that must be made. Generally, this work will occur during the period of time when verification work has already begun; thus we could choose to call it a change

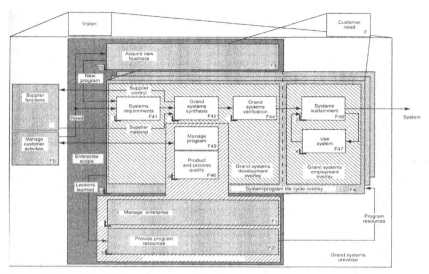

Figure 72 Reverification exposure periods.

in previously planned verification work or a reverification act. This may provide an opportunity for company and customer program managers to debate what to all it and how to deal with it from a program cost and schedule perspective.

So, the reverification impact will expand in time, including progressively more released items from the following checklist:

specifications	first article acceptance plans and procedures
analysis reports	qualification reports
engineering drawings	first article acceptance reports
simulations and models	system test and evaluation reports
computer software code lists	manufacturing planning
qualification plans and procedures	quality assurance planning
acceptance plans and procedures	technical data
system test and evaluation plans and procedures	training courses

A program can commonly cut off interest in modification and reverification before system use ends, because a user will, at some point, lose interest in spending money on an old system and will be thinking about replacing it with new resources. Modification of a system to achieve new capabilities is a major reason for reverification. Some extreme cases of this can be found in military aircraft programs.

The USAF/Boeing B-52 bomber was designed as a high-altitude long-range nuclear bomber but over many years evolved into expanded capability to penetrate a heavily defended air space at low altitude and high speed while retaining its original capabilities. It also evolved a capability to carry a tremendous iron bomb capacity and launch standoff weapons. The big ugly fat fellow (BUFF) also flew many tactical bombing missions during the Vietnam war followed by advancing clouds of destruction in the jungle below. Douglas Aircraft developed the C-47 as the DC-3 airliner, but World War II made many demands on this aircraft beyond what it was designed for. As late as the Vietnam war, the C-47 played a role as an attack aircraft featuring powerful guns shooting out ports in the port side of the aircraft while flying a circular flight path in a bank around a gun aim point, punishing a very durable airframe in ways never conceived by its design team while delivering a very punishing punch to the target. Many ground pounders developed a great deal of respect for "Puff the Magic Dragon" or "Spooky" when pinned down in the jungle. This mission was moved to the AC-130 Spectre for subsequent wars.

10.3.4 Process Redesign

Reverification is actually a process redesign activity. The understanding is that the process has been previously designed based on the content of the specifications at the time the verification work was originally planned. For one or more of the reasons addressed earlier, it has been found necessary to change some aspect of the verification process. At the time this kind of action is first understood, the verification planning is represented by a set of verification plans, procedures, and reports being managed using a set of matrices (verification traceability, verification compliance, verification task, and verification item) and a network diagram or combination of a process diagram and schedule, all derived ideally from data retained and maintained within a database system.

Therefore, any reverification work will first require some changes to this data. The question is, what is the impact of the proposed change on the existing data, any work that has already been accomplished, and future work not yet accomplished? The answers to this question are a function of when the decision is made to reverify relative to the sweep of program events involving verification. The term *reverification* implies that the verification action has already occurred previously, so we may assume for discussion that plans, procedures, and reports exist for the planned actions. Whether or not the plans and procedures have to be changed relates to the motive for reverification.

If the requirements and the corresponding product entity in question are different than they were when the verification action was first accomplished, then some subset of the planning, procedures, and reports will have to be changed based on a new set of one or more verification tasks. If the program previously built the verification planning data as covered in this book in Chapter 8 for qualification, Chapter 9 for acceptance, and Chapter 10 for system test and evaluation, then the same techniques for creating the planning data can be applied to edit it. This entails identifying the requirements that have changed, if any, and following those strings through the planning data to make any needed changes. This should reveal what tasks or parts of tasks must be rerun and reported upon. This may entail parts of all three classes of the verification process or only one or two classes. These tasks are replanned and scheduled, the work accomplished, reports created, and results audited with decisions made on the design relative to the changed requirements.

Reverification is one of the major reasons that the cost of correcting problems later in a program is greater than early in a program. Ideally, all of the verification work would have been properly executed based on sound planning and good requirements. Unfortunately, it often occurs that companies discover the quality of their early program system engineering work when the system enters item qualification, and the answer is that it was not done well. It is commonly less costly to follow the pattern of:

(1) define the problem in good specifications where every requirement is derived from a model and linked to verification requirements captured in the same document,

(2) solve the problem with cross-functional teams organized relative to the product entity structure and staffed with the right people, based on technologies that will be needed and risks that the program plans to avoid, and

(3) build the verification plans and procedures based on verification strings mapped to verification tasks that are placed under the leadership of specific persons.

CHAPTER 11

Enterprise Process Verification

11.1 IS THERE A DIFFERENCE?

The program requirements universe includes not only the product (or program-peculiar) specifications content but the process documentation as well. These requirements will be captured in documents called policies, plans, and procedures. Should we exempt these process requirements from validation and verification? It is suggested that the same situation is present here as in product requirements. Enterprises should have in place the machinery to determine the extent to which its programs comply with the contract, program planning documentation, and functional department manual content while accomplishing program work. This of course includes compliance with the cost and schedule parameters assigned to all program tasks.

It is possible to create process controls in such a way that they impose cost and schedule risks. It is very common that companies accept wildly unreasonable schedules offered in government requests for proposal (RFP) in order to get the contract. The author recalls that the advanced cruise missile program endured two proposal and selection cycles in the early 1980s. An initial operating capability (IOC) date was defined in the first cycle and it was very optimistic and risky. After many months of study and proposal preparation, the competing contractors turned in their proposals and the Air Force decided not to accept any of them. Employees of General Dynamics Convair reasoned that they had won according to the selection criteria and Boeing was supposed to have won. The Air Force put another RFP on the street with the same IOC as the first one and, after several more months, GD Convair was finally selected. Never has there been such a rapid transform between elation and depression as among the GD Convair team members upon hearing the news of the win. You could see it on the faces of the people who had survived two proposal cycles over a period of many months of overwork. First, the eyes opened wide amidst a face full of great satisfaction and an utterance, "We won!", followed almost immediately by a hollow look of despair and surrounded by the expanding inaudible wave, "Oh my God, we won." That program finally achieved its IOC quite a

325

bit later than the contract required, following a tremendous development effort that had to deal with multiple simultaneous technology challenges and overconstraining conditions, leading to one or more situations where the technical and programmatic solution space was nearly a null condition.

We should avoid creating program plans and schedules that have not been validated, that have not been tested for credibility. It is true that customers, as in the case just described, do not always offer prospective contractors good choices in terms of responding to their impossible schedule or losing the business. But when a contractor insists on responding to such requirements, they should offer a proposal that gives the best possible chance of success and identify the risks attendant to that plan, and where possible, a means to mitigate the related risks. The machinery for validation of these plans is risk management. You could argue that this is the case for product requirements validation as well, with few complaints from the author.

11.2 PROCESS VALIDATION

As we prepare the program plans, schedules, budgets, and procedures as well as the supplier statements of work and other planning data, the content should pass through the same kind of filter used for product requirements validation in our search for program requirements that can be achieved and that will achieve program goals. Whoever writes plan content should have a responsibility of including in those plans not only what is required in the way of work but the resources required, at some level of indenture, to accomplish that work (financial, time, personnel skills and numbers, material, equipment, facilities, etc.). Each task, at some level of indenture, should carry with it intended accomplishments and clear completion criteria telling how to identify when the task is complete. Realistic cost and schedule estimates, based on past performance, should be coordinated with these tasks. Satisfying these criteria encourages low risk planning, just as writing the verification requirements while writing the requirements in a specification results in better specifications. The purpose of validation, here as in the case of specifications, is risk identification and mitigation.

11.2.1 Completeness

Our process planning should be complete, free of voids. The best way to assure this is to create an overall process diagram that in Chapter 1 we referred to as a life-cycle functional flow diagram, an example of which is given new expression in Figure 73. For most companies, this diagram will

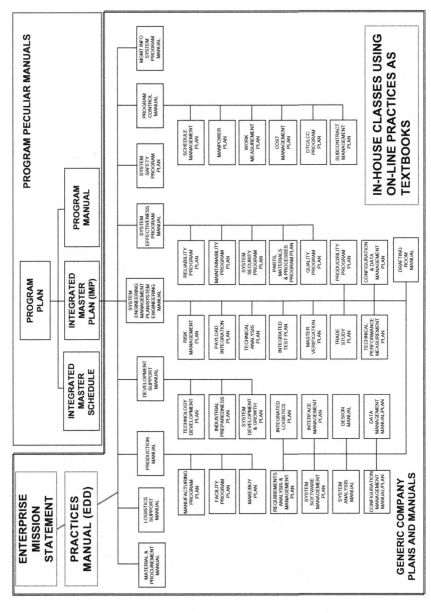

Figure 73 Program life-cycle functional flow diagram.

be appropriate for all programs in which they will ever become involved. The differences come into play at lower levels of indenture suggested in the diagram. The program uniqueness in this diagram primarily enters into the details of the product operational and logistics support functions appearing in function F47 where we may apply functional flow diagramming to expand our understanding of the needed functionality. The development, manufacturing, and verification (test) functions include program-peculiar content in terms of the specific product entities that must be developed but generally follow a similar pattern of behavior that can be depicted in a generic process flow diagram. In support of this largely generic process, over time, the author believes, a successful company can evolve a generic set of planning data that will require relatively small adjustments from program to program. One part of the planning information that should be nearly identical for all programs is the how-to planning.

In an organization that is managed through a functional staff supplying personnel to programs organized into teams in accordance with the system architecture and led through strong program management, this information can be supplied in the form of functional department manuals that provide the generic information on how to perform particular tasks given that they are required on programs. These manuals are supplemented on programs with program-specific plans that tell what work must be done, when it must be done, and by whom it must be done. The how-to coverage for all tasks called out in the program-peculiar planning is given in the functional department manuals, the whole content of which may not necessarily be implemented on any one program. On programs where it is necessary to deviate from the generic manuals, sections should be developed for a program-unique Program Directives manual that contains instructions specific to that program.

Company or division management should manage this program-unique manual building process by exception, offering criticism of the program manager where this manual takes exception to too large a volume of generic how-to content. The enterprise must encourage programs to apply the task repetition scenario to the maximum extent possible to maximize the continued drive toward perfection on the part of the staff. At the same time, it is through program experimentation that the enterprise will try new things and create new generic content. Once again, we run into an interest in a condition of balance between extremes. In the author's context, this is one of the things that an enterprise integration team (EIT), reporting to the top enterprise manager, would pay attention to. As discussed elsewhere in this book, the EIT provides the enterprise integration and optimization function adjudicating the conflicts across programs for functional management. Every organizational entity needs an integration and optimization agent and the EIT, by whatever name you prefer, is this agent for the whole enterprise, just as the program integration team (PIT) should be for each program.

During the proposal process, the EIT should formally review the planned program procedures and offer alternatives where the program attempts to run too far afield from the norm, yet encourage a few experiments in process. The danger is that too many programs will have in progress too many such initiatives, shredding the possibilities for common process repetition. Process change must be relentless but incremental and individually small. The program-peculiar excursions into process experimentation should be orchestrated based on past program lessons learned, derived from metrics put in place just for this purpose and evaluated for improvement priorities. New proposals and programs offer laboratories within which to experiment with new techniques. Otherwise, we may have to introduce changes while programs are in progress.

This arrangement encourages capability improvement through repetition and progressive incremental improvement through continuous process improvement of the generic department planning data, effective continuing training of the staff in new techniques learned through program experimentation, and improvement of the functional department tool boxes consistent with the practices. All of these resources are made available to all programs providing a foundation of excellence upon which to build the program-peculiar planning documentation.

Most aerospace divisions, companies, or business units deal with relatively few different customers and a relatively narrow product line permitting some heritage in the program-peculiar planning documentation as well. But, for any program, this planning data should expand within the context of a structured top-down process to encourage completeness. This process should start with a functionally derived architecture overlaid with a product work breakdown structure (WBS) augmented by process or services WBS elements that do not clearly map to any one product element. The WBS elements should then be expanded into statement of work (SOW) paragraphs giving high-level work definition. This is a relatively simple transform. For each product entity AX (where X is a string of base 60 alphanumeric characters) at some level of indenture we simply identify the following kinds of work statements:

AX_1 Perform requirements analysis yielding an item specification reviewed and approved at an item requirements review for item level or system requirements review (SRR) for system level as a precursor of design work. Where necessary, perform validation work to reduce risks identified in the transform between requirements and designs. Define verification requirements, plan verification work, and establish the verification management structure and database. Completion criteria is the release of the approved specification with customer approval where required or approval by in-house management at least one level higher than that immediately responsible for the specification.

AX_2 Accomplish synthesis of the requirements yielding a preliminary design reviewed and approved at a preliminary design review (PDR) and a detailed design reviewed and approved at a critical design review (CDR). Perform analyses necessary to support and confirm an adequate design. Completion criteria is release of at least 95% of all planned engineering drawings, release of all planned analysis reports, customer and/or company management approval of the CDR presentation, and closure of all action items related to the CDR.

AX_3 Perform integration, assembly, and test (IAT) actions to create the initial articles needed for engineering qualification analysis and test purposes. Complete all planned verification actions on the engineering articles, produce written evidence of compliance with the item performance specification requirements and present that evidence for customer and/or in-house management review and approval at a functional configuration audit (FCA). Closure criterion requires completion of all qualification tasks and related reports, customer acceptance of the FCA, and closure of all FCA action items.

AX_4 Manufacture a first article and subject it to acceptance verification actions driven by the content of the item detail specification. Present the results of the acceptance process of the first article at a physical configuration audit (PCA) and gain approval of the results from customer and/or company management. Closure criteria includes customer acceptance of PCA closure and closure of all PCA action items.

These paragraphs in the SOW should then be expanded into the content of an integrated master plan by identifying the contributing functional departments and obtaining detailed work definitions from each of them coordinated with the corresponding SOW paragraphs. The functional departments will bring into the plan particular intended accomplishments and associated detailed tasks that must be accomplished such that they can be mapped to the generic planning data content that tells how those tasks shall be accomplished. In an application of the U.S. Air Force integrated management system, the functional inputs to program or team tasks can be correlated with the significant accomplishments called for in that approach, as covered in the author's earlier book titled *System Engineering Planning and Enterprise Identity*.

This disciplined approach mirrors the same process discussed in Chapter 3 for product requirements analysis. And why not? The contents of the statement of work are the process requirements for a program. The author argues that the planning and procedures data correspond to the program design and this design should be responsive and traceable to the corresponding requirements. We first understand the grand problem and progressively expand that definition downward into more detail in an organized fashion. It is possible to create program planning data in a bottom-up direction but it is very difficult to integrate the resultant environment so as to encourage completeness.

For each of the tasks in the integrated master plan, we should identify the completion criteria, an appropriate budget and period of time in the program schedule when the task must be accomplished either in an event-driven or calendar-driven format, and its relationships with other tasks. By using a top-down disciplined planning process, there is a good chance that we will have satisfied the completeness criteria.

The documentation for the program planning strings could be included in hierarchically organized paper documents, placed in relational databases similar to those suggested for specifications and related verification data where the documents can be printed if necessary from the database records organized by paragraph, or organized into hypertext documents or multimedia data supplemented with video clips and voice. The latter approach, particularly, has a lot of possibilities for the generic how-to data. As noted previously, this information should be the basis for company recurrent training programs and in this format it could be used for just-in-time training of individuals or groups about to embark on a new program applying techniques for which individuals or groups need a refresher. This information would serve equally well, coupled with a good find function, as a look-up source on the job for specific concerns.

Figure 74 offers one possible view of the aggregate enterprise documentation approach. The enterprise mission statement has been expanded into a

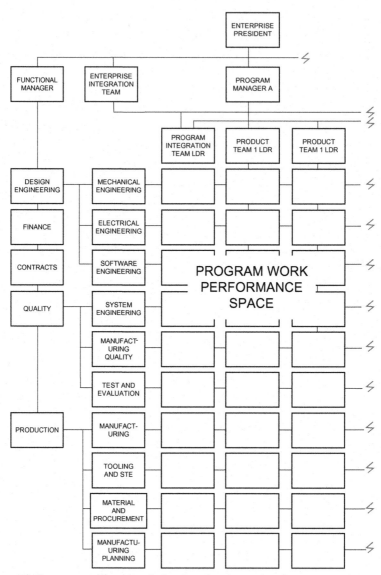

Figure 74 Program and functional planning documentation tree.

high-level practices manual giving global requirements defined by top management, including the functional organization structure definition and charters for those organizations. All of the how-to documentation expands from this base under functional management responsibility. The program planning documentation is expanded from a customer need expressed in a top-level product architecture defined though a careful mission analysis overlaid by a product and process work breakdown structure (WBS), expanded in turn into a statement of work (not shown in either case). The work defined in the SOW is summarized into a simple, top level program plan and further expanded into an integrated master plan expressed also in terms of timing in an integrated master schedule.

The program manual included in Figure 74 offers the program-peculiar procedures discussed previously. In general, this manual simply refers to the functional manuals as the source of detailed instructions in all cases where an exception is not presented in some detail in the Program Manual. Integrated master plan content refers to the generic how-to manuals for details, unless an exception is included in the Program Manual, in which case the alternate procedure should be included in total.

Figure 74 also encourages the use of the generic how-to manuals as a source of textbooks for in-house courses. These may be delivered as organized one-day or multiple-day classes, fragmented in lunch time classes, self taught via the network in a just-in-time fashion, or simply read by the practitioners in paper form. Your company may not have the time or money to devote to generating the detailed information required to support the textbook application. An alternative is to prepare the plans so as to describe what to do and the sequence in which these activities should be done, without providing great detail in the methods of accomplishing the work. To complete the story, you can refer in each section to standards, guides, or textbooks that do cover the way you prefer to do the work. The result is the same as if your company had devoted the time to write all of this material. In those few cases where you cannot find a suitable reference, you may have to prepare a preferred procedure.

11.2.2 Accounting for Planning Risks

As the content of our SOWs and program plans flows into reality, it should be screened for risks at some level of indenture, just as we would perform validation on the content of our specifications as we create the content. Most of the risks that may become embedded in our planning documents will be of the cost or schedule variety rather than product performance, which are primarily driven by specifications content relative to our company's design capabilities and the technologies available to us.

We must ask whether the content of our plans is realistic in terms of the amount of budget and schedule available. Given that we have identified all of the tasks that must be accomplished at some level of indenture, budget and time should be allocated to them and combined in a process cost and schedule math model so that we can manipulate the model in a what–if fashion. There are many available project-planning software packages that are very effective in this application. Most large companies use one of these tools for manufacturing planning but may not make it available for more general planning purposes. If all else fails, adequate software of this kind can be acquired for operation on Mac and Windows type desktop computers.

Margins should be included for cost figures at some level of indenture as a function of our perceived risk and the nature of the contract cost structure (cost plus incentive fee, fixed price, etc.) and float added to task schedule figures on the critical path as a function of risk as well. The resultant cost and schedule totals should be compared with what we know or think to be consistent with customer expectations and what we think to be a competitive position. If our cost or schedule figures are in excess of these, then we have to make a decision about whether to try to reallocate to close the gap or stick with our position and, if necessary, refuse to make a bid on the program because it poses too great a risk. In the former case, our proposal would have to include a clear and unmistakable statement of the program risk attendant to the customer's original cost and schedule figures in order to ensure that the customer fairly considers its proposal amidst the other proposals of other bidders that may accept the risky position as a means to get in the door. Government customers are becoming more sophisticated than in years past when it comes to identification of risky proposals and will more often today give a truthfully defined program with attendant risks than many have observed from past experience.

It is possible to generate cost and schedule figures from the bottom up, also referred to as a grassroots estimate. The result can be the same as discussed here so long as we compute the totals and make the comparison with expectations and make the adjustments suggested by the comparison or decide not to bid. The danger in the bottom–up estimate and method of building the program task structure is the greater potential for omission of needed tasks. The top–down method encourages completeness. At the same time, the top–down decomposition method does run some risk of disconnecting important interfaces between tasks, as is the case in all applications of the decomposition scenario. Therefore, integration and optimization of the expanding task structure is necessary just as it is in the results of a functional decomposition applied to the product.

The author has in mind a simple functional flow method of decomposition as suggested in Figure 73 but sees no fundamental flaw in applying an

MSA-PSARE, UML-SysML, or UPDM approach, to name a few. All of these methods provide an environment within which we humans can seek to understand complexity in an organized fashion encouraging completeness.

The way estimates are commonly created in aerospace companies dealing with the government is that a grassroots input is requested from the contributing departments or functions. Seasoned veterans of this process build in a hidden man-hour margin because they know that management will skim a management reserve. This is all a charade and a very poor example of sound management techniques. A program organization is better served to encourage honest estimates and review the inputs and margin figures for credibility. On a program that is going to be managed by cross functional teams, the task estimates should be built around the planned teams with inputs offered by the participants from the several participating functional departments and integrated by team leadership into an optimum estimate with margins and floats driven by perceived risks.

The aggregate margin and float figures become the program reserve controlled by the program office with some portion perhaps allocated to product team control. During program performance, the teams do their very best to accomplish their work without appealing to their margin and float. Depending on the risks that materialize, program and team management should allocate margin and float to resolve problems that cannot be resolved by other means. The cost figures and schedule demands included in the original estimates were based on perceived risks at the time they were determined, and over the run of the program some of those will materialize while others do not and still others never conceived will also come into play. This results in a mismatch between the basis for the margins and floats and the reality of program evolution but it is the best that can be done at the time the margins and floats are determined, because you are dealing with the future. The pluses and minuses will likely even out over the program duration and the aggregate margin and float will be consumed, though not in precisely the same way it was created.

This approach is identical to the use of margins adopted for the product requirements as a risk management technique. The requirements that present the greatest implementation risk are assigned margins and over the program development life these margins are generally consumed in the process of responding to risks that evolve. They commonly are not consumed in precisely the way they were originated. In some cases, a margin in one area such as weight or cost may be cashed in to improve the reliability of an item that is possible with a little more weight or a little higher design-to-cost figure. This is all part of the system optimization activity and it can be applied to process as well as product. What is more, the notion of margin class interchange just noted can be extended to the margin and float figures for process. That reliability problem might also have been solved if the schedule could be extended slightly (with other possible impacts) or if the budget could be expanded slightly.

In the latter case, we see a marriage between the design to cost (DTC) product requirements and the use of program task budget margins. The task budget margin very likely will provide the resources for the DTC program. It is important to manage the complete budget and float structure together rather than having engineering run a product requirements margin system and finance run a cost and schedule margin and float system. Whatever agent is globally responsible for program risk management, and in the author's view this would be the PIT, should be responsible for managing all of these margin systems.

11.3 PROGRAM PROCESS DESIGN

As in the case of product design, the process design should be responsive to the process requirements captured in our program planning documentation. The process design consists of detailed procedures to implement planning, personnel selection to satisfy predefined specialty engineering needs and establishment of the means to acquire these people, definition and acquisition of the program space or facility and association of the personnel and teams with this space, identification of the needed material resources and acquisition of them, clear assignment of previously defined responsibilities to persons or teams, and development of the information and communications resource infrastructure. Upon completion of the process design work we should be ready to light the program pilot light and turn up the gas. We should be ready for program implementation in accordance with a sound validated plan.

11.4 PROCESS VERIFICATION

The verification of our process plans and procedures is accomplished in the implementation of them by monitoring preplanned program and functional metrics and using the results as the basis for actions taken to improve program performance. As in the case of product verification, process verification occurs subsequent to the design and implementation phase. There is a difference in that the results of the process verification work may be more useful on the next program rather than the current one. Metrics collected early in the program may yield improvement suggestions that can be applied prior to the end of the program but metrics data collected later in the program may not be useful for that program. This should not lead the reader to conclude that program process verification is therefore a worthless waste of time and money. Quite the contrary, it is the connecting link for your continuous process improvement activities that will, over time, progressively improve company capability and performance. It provides a conduit for lessons learned from each program experience to the company's future of diminished faults.

11.4.1 Program and Functional Metrics

Metrics are measurements that permit us to measure our performance numerically and therefore make comparisons between programs and within programs at different times. Given that we have a generic process that can be applied to specific programs, that we have generic how-to documentation in the form of functional department manuals for all of the tasks expressed on the generic process diagram as blocks, and that we actually follow these practices on programs, we have but to determine where to install test points in our process diagram from which metric data will flow. A good place to start is to list the key tasks in our process life cycle, such as the representative systems engineering tasks listed in Table 10.

This list is not intended to be comprehensive. The SENSE column tells which value sense is desirable. Refer to the *Metrics Guidebook for Integrated Systems and Product Development* published by the International Council on Systems Engineering (INCOSE) for a good source of specific metrics related to the systems engineering process.

The metrics selected should coordinate with specific tasks that are generally applied on all programs. Unusually good variations from expected values on specific programs should be investigated for cause as a potential source of process improvements, whereas the opposite should be investigated for opportunities to improve the performance of specific personnel, reevaluate resource allocations, or correct personnel performance. Poor performance repeated from program to program should encourage us to select that task as a candidate for process improvement.

It is assumed that the reader is familiar with how to track these measurements in time graphically and to use the display of performance as a basis for

Table 10 Metrics List Representative Candidates

Task name	Metric candidates	Sense
Proposal Development	Red Team Quality Rating	High
	Cost of Preparation as a Fraction of the Program Cost Estimate	Low
Mission Analysis	Man-hours Required	Low
	Number of Requirements Derived	High
Functional Analysis and Decomposition	Number of Requirements Derived	High
Requirements Analysis and Validation	Average Man-hours per Specification	Low
Specification Management	Number of Requirements Changed Subsequent to Specification Release	Low
	Average Number of Requirements per Paragraph	Low
Requirements Verification	Average Number of Verification Tasks per Item	Low
Risk Management	Program Risk Figure at PDR	Low
	Number of Active Risks	Low
	Number of Risks Mitigated to Null	High
Major Review Performance	Average Number of Action Items Received per Review	Low
	Average Number of Days between Review End and Review Closure	Low

management action to improve performance. Ideally, each organizational entity would have a series of metrics that are a good basis for determining how well their process is doing. The process-related measurements will likely relate to cost and schedule performance primarily, but may include other parameters. They should also be held accountable for a subset of the product metrics most often referred to as technical performance measurements discussed in Chapter 1.

11.4.2 Use of Cost/Schedule Control System in Process Verification

For each step in the program process, at some level of indenture, there should have been allocated appropriate schedule and budget through the WBS associated with the step. All program work should roll up through the IMP and SOW to the WBS. The contractor's cost/schedule control system (C/SCS) should produce reports telling the current cost and schedule status for each task at some level of indenture. The histories of these parameters should be retained and used as a basis for adjusting estimates for future programs. The C/SCS can become one element in a closed loop control process driving the cost estimates toward low risk values. This is a hard process to start because it only offers a delayed gratification rather than an immediate payoff. This same problem stands in the way of many improvements. The solution to this impediment is to simply accept that we will be in business for some time so we should work toward the long-term benefit of our enterprise.

11.4.3 Progressive Planning Improvements

The metrics data collected from programs should, over time, lead us to conclusions about tasks that our business unit does very well and those that are done very poorly. The latter lessons learned should be the object of study to determine how we can improve performance on future programs. First we need to determine what tasks are done poorly and then we need to prioritize the corrective actions needed to improve performance. These actions may entail improving our written practices (generic or program-peculiar), computer tools and other supporting resources, facilities, or the training or skills base of our personnel. We need to determine which combination of these approaches would be most effective in correcting the high-priority problems determined from evaluation of our metrics collected on programs.

11.5 ORGANIZATIONAL POSSIBILITIES

Up to this point in his career, the author's organizational framework has been a company with multiple programs organized into cross-functional teams with personnel and resources supplied by a traditional functional management structure. The author's past also has prejudiced his perspective in the direction of a strong independent test and evaluation organization functionally organized and run across several programs. His preferred model also includes an independent quality organization and a systems engineering function under engineering. In many companies, engineering management has been incapable of optimizing the application of the powerful force called system engineering to positively influence the product and process including the performance of engineering itself. The possibility exists that there may be a better way to relate these three organizational functions unveiled through an interest in verification. The author feels that even more could be said to support this idea, so will burden the reader with additional comments about this possibility.

Let us first extend the word *inspection* to embrace all five validation and verification methods we have discussed (test, analysis, demonstration, examination, and special) in keeping with MIL-STD-961E usage. Granted, some would use the term *test and evaluation* to extend across all five methods and the author has no quarrel with that term either. The application of the inspection process extends across the full length of the development and manufacturing process. Early in the development we apply it to validate the transform between requirements and product, giving us confidence that we can create a compliant design. Subsequent to the design process, we manufacture product elements and subject them to qualification inspections, largely through the work of a test and evaluation function within an engineering function, proving that the design does satisfy the performance requirements. During production we subject the product to acceptance inspections to prove that the product properly represents the engineering and manufacturing product definition. The acceptance testing work is commonly staffed by manufacturing personnel and overseen by quality assurance persons.

Through all of this work there are those who do the related work and those who should coordinate, orchestrate, integrate, or manage it, and those who assure that it has been correctly accomplished. Table 11 lists all of the validation and verification possibilities and identifies the party who, in the author's view, should be responsible for coordinating or managing the work, quality assurance, and who should be responsible for implementing the indicated work.

Manufacturing quality assurance is but one application of the verification process. It may make for a more efficient and technically sound process to scoop up the current test and evaluation and quality organizations into a validation and verification (V&V) department by whatever name. The author has concluded from observation of the practice of systems engineering in many organizations that the engineering functional organization and systems engineering should perhaps have a relationship similar to that of manufacturing and quality assurance. When quality was under the manufacturing organization in many companies, it was compromised and unable to assertively counter management insistence on production over quality. By separating the function, it was empowered to participate on an equal footing.

It is too often the case that persons arriving at positions of responsibility for engineering organizations never had or mysteriously lose the good sense to encourage the systems approach in the organization they inherit. Efforts to make improvements from below become career-limiting, resulting in

Table 11 Validation and Verification Responsibilities

Level 1 Task	Level 2 Task	Method	Coordinator	Quality Agent	Implementation
Validation	–	Test (DET)	T&E	Quality	T&E
	–	Analysis	Domain	Quality	As Assigned
	–	Demonstration	Domain	Quality	As Assigned
	–	Examination	Domain	Quality	As Assigned
		Special	Domain	Quality	
Verification	PMP	All	Domain	Quality	Domain
	Qualification	Test	T&E	Quality	T&E
	Analysis	Domain/Team		Quality	As Assigned
	Demonstration	Domain/Team		Quality	As Assigned
	Examination	Domain/Team		Quality	As Assigned
	Special				
Acceptance	Test	T&E		Quality	Manufacturing
	Analysis	T&E		Quality	Manufacturing
	Demonstration	T&E		Quality	Manufacturing
	Examination	T&E		Quality	Manufacturing
	Special				
Systems	Test	T&E		Quality	T&E
	Analysis				
	Demonstration				
	Examination				
	Special				

burnout for those who try. The organization suffers because it never places an emphasis on quality systems engineering and it never achieves it. Figure 74 offers an alternative that will please no one except quality engineers, perhaps. It places the traditional systems engineering, quality assurance, and test and evaluation functions together in one functional organization called Quality. The reader can pick a name more pleasing to him or her if desired. Persons who have come to love the term *total quality management* might select that term as the title of this organization and finally see it achieved.

The systems function in Figure 75, repeated from Figure 74, is responsible for risk management, mission analysis, system requirements analysis and management, specialty engineering (reliability, maintainability, logistics, safety, etc.), system analysis (aerodynamics, thermodynamics, mass properties, etc.), materials and properties, parts engineering, integration and interface development, system engineering administration (reviews, audits, action items, etc.), configuration management, and data management. Manufacturing quality is the traditional quality assurance organization responsible for verifying the quality of the process and product. The author would extend this responsibility to the complete process, such that all functional organizations contributed practices, tools, and qualified personnel to programs and insist that their performance be audited on those programs in accordance with the written practices using metrics selected for that purpose. The new aggregate quality organization should have global process quality responsibility.

The test and evaluation organization is responsible for the management of the product V&V process as described in this book, as well as providing test laboratories, resources, and test engineering skills appropriate to the company's product line. People from this organization manage the process using the verification compliance and task matrices; they schedule the whole process, energize the responsible parties and resources at the appropriate time using planned budget, and track performance/status throughout. Someone from this functional organization assigned to the PIT would lead the whole verification process composed of test, analysis, demonstration, and examination activities assigning responsibilities as appropriate.

The engineering organization in this structure is simply the creative engineering organization that solves problems defined by requirements in specifications. The results of their work in synthesizing the requirements is evaluated by the quality organization through validation work primarily driven by engineering concerns, verification work leading to qualification, and verification work leading to acceptance of specific product entities. The kinds of engineering specialties required are, of course, a function of the product line.

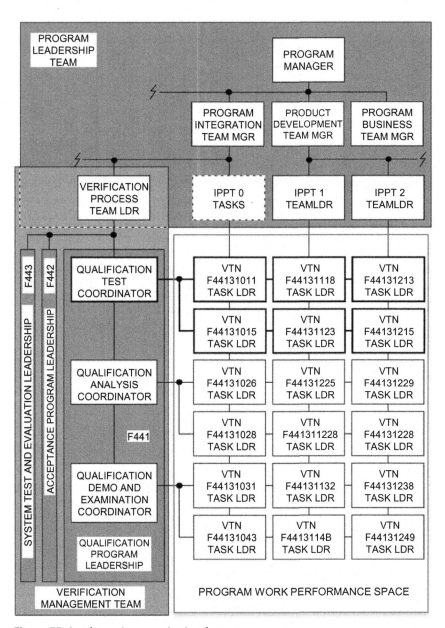

Figure 75 An alternative organizational structure.

The quality organization in the aggregate provides the program framework within which the people from all of the indicated organizations perform their work on each program in what is referred to in Figure 75 as the program work performance space. Within this space all of the people assigned to the program from the functional organizations are organized into teams oriented toward the product architecture for reasons exposed in Chapter 1. Only one of potentially several programs is illustrated.

The enterprise integration team works through the program integration teams to integrate and optimize at the enterprise level identifying conflicts between programs and working to adjudicate a good solution. A principal example of this is isolated program planning that calls for simultaneous use of enterprise resources beyond their current capability. One or more programs may have staffing peaks that exceed enterprise personnel availability and it may be possible to level some of these peaks by shifting program goals in time without violating customer interests. Similarly, multiple programs may call for the same test or manufacturing capabilities in the same time frame. The EIT should also reach across the programs to encourage commonality of process by working with both program and functional organizational structures and managers.

CHAPTER 12

Postscript

12.1 POSTSCRIPT PLAN

In the prior chapters we have explored the V words, offered clear (but not necessarily universally accepted) definitions, described the related work that must be done, and offered documentation templates for corresponding planning and reports. Generally, the content has recognized the system engineering and management concepts extant at the time the book was rewritten in 2015. While rewriting the book, however, the author reached some conclusions about the status quo that he would prefer to share with others. These matters have a bearing on the performance of verification work, but some of them also spread out over the whole systems approach as well.

12.2 CLOSURE ON MEANINGS

Hopefully, this book will bring attention to some of the vocabulary differences between hardware and software people and encourage closure toward a common vocabulary for systems. The development of products to satisfy our very complex needs continues to be a very difficult process and promises to become more so rather than less, because our past successes encourage us to take on more and more complex problems. As we solve difficult problems, we become aware of the fact that our past solutions were but subsets of grander problems, encouraging us to merge multiple systems by automating their interfaces. In order to avoid development-retarding forces, we should move to reach agreement on a common vocabulary across the widening hardware/software abyss. The decision as to which terms are favored and which are dropped is of absolutely no consequence. This is true of the V&V words that this book focuses upon, as well as all of the other terms in conflict. As noted in the foreword, this book was at least in part motivated by frustration over the different meanings assigned to these two words and it has been a fulfilling experience for the author to express his opinions on this matter. But it would be far better for us all to agree on the meaning of these words, even if it happened that we selected the opposite or a different meaning from those expressed in this book.

System Verification
http://dx.doi.org/10.1016/B978-0-12-804221-2.00012-7
345

12.3 HOPES FOR BALANCED TREATMENT

In the past, the existence of strong functionally oriented test and evaluation departments in industry and the corresponding investment in costly test laboratories requiring good centralized management have encouraged the evolution of a mindset emphasizing test methods regardless of the situation. This is a positive trend, but the other methods commonly have no single unifying functional mentor. It is hoped that the continuing interest in the cross-functional team concept will relax the relative strength of the lobby for test versus other methods. The author is an advocate of matrix management with strong programs organized by cross-functional teams oriented about product entities and supportive functional departments providing programs with the resources they require (qualified people, tools, and generic practices, as well as test laboratories). So the answer is not to reduce the excellence that test and evaluation organizations have achieved, but to find a way to give the other methods a home that is every bit as effective in promoting excellence.

An organizational improvement that could be considered is to give the traditional functional test and evaluation department the full verification management responsibility, perhaps renaming the department Verification. It is not suggested that personnel of such a department actually accomplish all verification work. Reliability engineers are going to have to perform reliability analyses, structural analysts will have to perform their specialized work, and logistics engineers will have to continue to perform maintenance demonstrations. What is suggested is that some of the people we now call test engineers become more broadly qualified to embrace skills needed to understand and coordinate verification work under all five methods. On a given program, we would find these kinds of engineers on each product team and on the system team. The verification engineer would orchestrate the verification process for items and the system as discussed in this book. Where test is appropriate, a pure test engineer would be assigned the responsibility to plan and manage the work, possibly using a test laboratory managed by the functional Verification Department and made available to all programs. Verification work corresponding to other methods would be similarly assigned to other experts in the appropriate field.

People assigned to programs from the Verification Department would be responsible for management of all of the work embracing Section 4 of all of the specifications, verification task planning, and implementation of this work. These people would act as project engineers working from within the system team to orchestrate the complete process across the program. Most of the verification work would be accomplished by specialists within the teams, but their performance would be guided by the verification documentation described in earlier chapters to which they may have to contribute. On a large program a verification process leader in the PIT may have reporting to him or her qualification, acceptance, and system verification leaders. Each in turn manages strings of task leaders through method coordinators, as shown in Figure 75. On smaller programs, much of this overhead can be collapsed into one overall verification leader/coordinator.

This suggestion is in keeping with the attitudes that some test and evaluation engineers have that their activity is really the super-set of verification and that the test and evaluation boundary is an overly constraining influence. Most other solutions, including continued dispersal, are not constructive. It makes no sense to create an analysis functional department or a demonstration department. Demonstration, examination, simulation, and analysis tasks are appropriately done by people from many different disciplines and those disciplines correspond to the correct orientation for the functional organizational structure. Test should be centralized within a company because of the need to manage the test laboratory resources efficiently. The other four methods are appropriately distributed across all other disciplines, but the work they do on programs should have central management or coordination on programs.

A more radical organizational change would involve teaming the traditional system engineering, test and evaluation, and quality assurance organizations into a single functional department. The grouping of test and evaluation and quality might be more easily achieved as a first step on the road to achieving a true total quality organization. The engineering organization may offer a lot of resistance to separation of the systems function from their control, even though in so many companies engineering management has failed to encourage the growth and effectiveness of its system function. The author believes that it is high time that we tried some other alternatives enroute toward system engineering excellence.

Many professional system engineers and managers may feel the combination of systems engineering, test and evaluation, and quality functional departments attenuates the role of the system engineer in the development process, but the reality is quite the contrary. It could expand the role by accepting test and evaluation engineers as system engineers and associating those who retain the job title of system engineer primarily with the development phase of a program, beginning when there is only a customer need and moving through the program activity until critical design review (CDR). There is a rich tapestry of work between these events that is broader than most of us can support. There will still be system engineers who focus on mission definition and others more proficient in engineering and manufacturing development. There is also and will always be a need for the people Eberhardt Rechtin calls system architects, people the author is comfortable referring to as grand system engineers, who are system engineers working at a higher plane of abstraction with very broad lateral boundaries.

Subsequent to CDR, the design should be relatively static, yielding ideally only to engineering changes encouraged by the customer based on their improved understanding of the solution to their problem and the new possibilities that evolving product capabilities encourage. Most post-CDR work on a well-run program should be focused on integration, assembly, and test of real hardware and software. The tough development choices should have been made long before and the program should have a sound foundation upon which to manufacture and code the product confidently. Yes, there is system engineering work remaining to be accomplished but the verification engineers we have described would be well suited to do that system engineering work.

The author has drawn criticism from fellow system engineers in the classroom and work environments about associating the quality assurance function with system engineering. The criticism commonly offered is that Quality is generally staffed by less-qualified engineers and non-degreed persons, and that associating closely with Quality engineers will somehow bring down property values and "ruin the neighborhood." If this is true, then that Quality organization is not staffed with quality people, which identifies a separate problem which industry should be doing everything possible to resolve.

There is a much bigger problem in the application of a quality system engineering process in many organizations than others dragging the good name of system engineering down, however, if that premise is true. There are companies with a strong and respected system capability, but in many companies it would be hard to imagine the system engineering function sinking any lower in esteem within the organization. Too often, engineering management is less than enthusiastic about the system function. Functional managers are allowed to interfere with or even lead the daily workings of programs leading to serial work performance on programs. All too often, the systems charter has become a paperwork-dominated activity responsible for performing the residual work that no one in the design and analysis functions wants to do, such as documentation. The difficult product technical integration work either goes undone or it happens by chance between the design engineers who happen to stumble upon conflicts. The same criticism directed by some at Quality engineers is often valid for system engineers in some companies in that, over time, engineers who can't make it in design or analysis collect in the paperwork function.

Many engineers and engineering managers reacted very negatively to the requirements of Air Force Systems Command Manual (AFSCM) 375 series in the 1960s, which required a very structured process complete with a specific set of forms coordinated with key punch data entry from the forms into a mainframe computer. Each intercontinental ballistic missile program offered a little bit different version of this manual through the 1970s and 1980s, but they were all dominated by 80 column Hollerith card technology for mainframe computer input. Many engineers who experienced this environment despised the structure and the tedium associated with these processes. Some of these engineers matured into the engineering managers and chief engineers of the 1980s and 1990s, and in many of the organizations where these standards were applied by these managers, the work was relegated to people called system engineers focused on documentation. The logical, artistic, and technical skills required for requirements analysis and technical integration fell into disuse in many of these companies.

While supporting a company that was developing a proposal for a space defense system, the author commented that they had used a hierarchical functional analysis process to define the system architecture and that it was likely that they had not clearly understood all of the needed functionality nor concluded with the optimum architecture. It was proposed that a functional flow approach be tried to determine if a simpler system might result. The proposal manager did not want to impede proposal progress, so he asked the author and one of his own company's system engineers to do that analysis in another city where their systems division was located. The analysis was done in a very few days, yielding some ways to simplify the deployment system significantly by optimizing at a higher level of integration including the launch vehicle, but the proposal was not going to be changed at that point. The company did not win and should not have won that competition.

The author heard the same refrain from several chief engineers and proposal technical managers over the period of the 1980s: "I don't want my creative engineers damaged by that documentation mess." Other expressions of this feeling included, "System requirements analysis only consumes a lot of trees"; "SRA has no beginning and no end, just an expensive middle"; and

"Monkeys could fill out these forms and that's what I feel like while doing it."

More recently, companies sold on cross-functional teams eliminate their functional organizations, including system engineering, and projectize, thereby losing the glue that holds the whole continuous improvement machinery together. Some otherwise gifted managers believe that it is possible to assemble cross-functional teams that will flawlessly interact to develop common interfaces without the need of a system integration and optimization function. Apparently they feel that somehow people will magically start talking to one another simply because of an organizational change, where this never happened before while the company was a matrix with strong functional management and programs dominated by serial work performance, with all of its related problems.

Aerospace companies know that they must put on a good show in the system engineering area because their customers have come to respect the value that a sound system engineering activity should add to their product, even though the members of the customer program office team may not clearly understand how this magic is supposed to happen. Too often, this is mostly show with little substance, and the customer is unable to detect the difference. There are too many stories, like the company that was developing a very advanced aircraft without delivering the required system engineering data items. After repeated warnings from the customer, the company hired some system engineers and assigned them the required work separated from the *real* engineers who were continuing with the design effort (although in the same town, in this case). Once the data was completed and shown to an appreciative customer, the system engineers were separated and the program continued as before.

There is a lot of real system work to be done in companies and on programs, but the current relationship between design engineering and system engineering management is not optimum. The system engineer needs to be set free from the constraining power of design engineering management in order to help the whole organization, including design engineering, move toward perfection. System engineering needs a strong voice outside the design engineering channel supporting its proper activities on programs as the product system technical integration and optimization agent. In doing this work, it is normal to offer constructive criticism of the design and analysis functions and their ongoing work. Those who must offer this criticism should not have to work within a regime where the top management of the design function has career power over them. An alternative is the selection of persons to manage engineering who have a broader view and an understanding of the fundamentals that drive specialization and need for system engineering.

Some examples of similar conditions that make it difficult to accomplish one's work with integrity include:

a. The quality assurance engineer whose reporting path includes manufacturing management.

b. Companies that manufacture commercial aircraft are required to have people on the staff called designated engineering representatives (DER). These people are on the company payroll but are responsible for assuring compliance with FAA regulations. In order to satisfy their function, they have to have a large degree of immunity from company management retribution despite the very uncomfortable message that they may have to bring to the table from time to time.

c. While the author was a field engineer for Teledyne Ryan Aeronautical in the 1960s and 1970s, he often experienced a case of divided loyalties while in the field representing his company's product. The ultimate customer (Strategic Air Command for most of the author's field service work) was always rightly interested in the truth about the product and this was generally, but not always, the attitude of the company and the government acquisition agent. The author's conclusion was that no matter how currently painful a piece of information might be, immediate truth was a better selection in the long run. The results of this course of action will not always be appreciated on all sides. For example, the author found that another field engineer had told the troops how to adjust the Doppler pulse scale factor potentiometer based on prior mission inaccuracy to improve future mission guidance accuracy. There was no technical data to cover this adjustment and there was no guarantee that others who were briefed on the procedure verbally in the absence of the field engineer would use this capability correctly. One night while the author was socializing with the location commander, he told the colonel what was going on and asked him to avoid the military solution. The next day the colonel applied the military solution, making everyone else very angry with the author for spilling the beans.

One might conclude that the system engineer needs to be freed from these kinds of constraints so as to be able to contribute critically and forcefully to the evolving preferred solution. There is a good counter-argument to this, of course: that every engineer should simply act on their own opinion of the best choice of alternatives and take the benefits and hits where appropriate.

12.4 TOOLS

Few companies have taken full advantage of available computer technology to improve their verification capabilities. As discussed in earlier chapters, all of the content of all of the verification documents can be captured in relational databases providing traceability throughout. On-line data in database format from which documents can be created as needed used in combination with computer projection for meetings can have an explosively positive effect on the velocity with which one can accomplish this work and offers a significant cost reduction opportunity, eliminating related reproduction, distribution, and lost time or inefficiency costs.

Most companies have already made the investment in machinery (computers, servers, networks, and software) and have only to creatively apply those assets to this very significant information management problem.

Unfortunately, the tool companies have not yet expanded their interest to embrace the full verification problem. The author believes there is an unfulfilled market available to the tool maker that opens their eyes to a broader view. It will not be sufficient to simply build the tables corresponding to the needed data content. The successful toolset will permit distributed operation with assignment of fragments to particular persons or teams and central management of the expanding information base. These tools should also take advantage of some of the creative ideas put into play on some programs permitting access to the data in read-only mode via their intranet by a broad range of people who are not required to maintain skills in several tools, each of which can be quite difficult to master and retain skill. There is a need for tool integration and optimization at a higher level of organization.

12.5 SINGLING UP THE WBS LINES

The good ship WBS (work breakdown structure) has been secured to the dock by two lines in the past, the product and the process (or services) components. Commonly, test and evaluation, a principal part of the verification process, is a separate services-oriented WBS. The author maintains this is unnecessary on development and production programs and that the added complexity is an impediment to effective and efficient program planning. The whole WBS for a development or production program should be product oriented in that services are accomplished in all cases to develop product at some level of indenture. The product and the process used to produce the product are two very different things, deserving unique orthogonal identification.

Customers wish to have a product delivered. He who would create and deliver that product will have to expend work on every element of that product, thus providing services. Today, the common practice is that work that can be easily correlated with the product entities will be scored against the WBS for those product entities. The work that cannot easily be associated with any of the entities selected as WBS elements or which must be associated with the system level will be scored against services WBS elements, such as program management and system engineering.

If we recognize that work can be accomplished against the whole system as well as any element of the system, we shall have a workable model that will enable an efficient transform between generic enterprise work planning and the work needed for a specific program. The DoD customer has, in the past, been so concerned with their need to control cost and schedule that they have forced contractors to apply a planning model that denies them an efficient map or transform between their identity and the needed program work. Therefore, these contractors have generally failed to develop an identity through common written practices to be applied on all programs, sound internal education programs (on-the-job training and mentoring accepted as well as more formal methods) focused on that generic process, and efficient methods for building program planning structures based on their process definition.

Figure 76 illustrates the proposed relationships between product items, functional departments, generic work commonly required on the enterprise's programs, and time. The understanding is that the enterprise is managed through a matrix with lean functional departments responsible for providing programs with qualified people skilled in standard practices found to be effective in the company's product line and customer base and good tools matched to the practices and people skills. Each of these functional departments has an approved charter of work they are responsible for performing on programs and which provides the basis for personnel acquisition and training.

The front vertical plane in Figure 76 illustrates the aggregate functional department charter by showing the correlation between all functional departments and all possible work they may be called upon to do on programs. The darkened blocks indicate this correlation. The crosshatched block in each column indicates the department that should supply the persons to lead that particular effort on programs in the product development teams assembled about the product hierarchy. The indicated departments must prepare practices supportive of the practice defined by the task lead

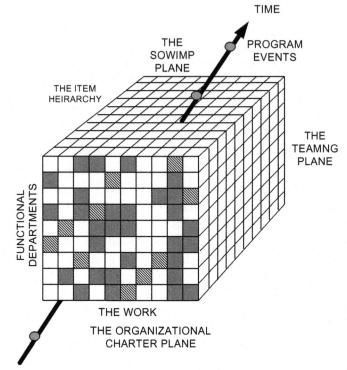

Figure 76 Product-only WBS enables good planning.

department (crosshatched block). The task lead department should also be primary in providing tools to do the work for that task on all programs and for providing any related training.

The reader should note that when an enterprise fully projectizes its work force, it loses the agent that can accomplish the generic work useful in providing the foundation for all programs. Each program becomes a new process factory. Therefore, there can be no central process, there can be no continuous improvement, there can be no progress – only a never-ending struggle, at best, at the lower fringes of excellence driven by the superhuman efforts of one or more heroes.

In the suggested arrangement, the WBS and the product hierarchy become one, since there is no process component of the WBS. When planning a program composed of items identified on the item hierarchy (or WBS) axis in Figure 77, we must select the work required from our generic work axis forming the statement of work (SOW) and integrated

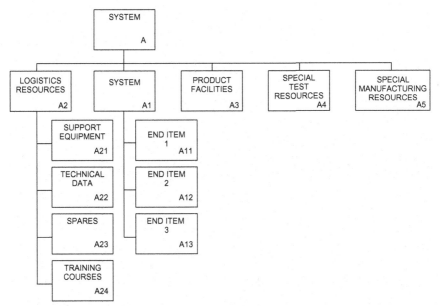

Figure 77 The product-only WBS.

master plan (IMP) content (the SOW/IMP plane). This work selection leads to a clear identification of the functional departments from which personnel should be selected and the practices that must be applied on the program. Once we have decided the product entity nodes about which we will form cross-functional teams, the teaming plane tells us to which teams these people from the functional departments should be assigned.

The SOW that results from the first level of correlation between the product entity structure and generic work must then be expanded into a more detailed plan. For each SOW paragraph, we need to identify in a sub-paragraph in the IMP the work to be accomplished by each contributing department, which should include the cross-functional, co-process integration work to be accomplished by the task lead department. All of the work thus selected must then be placed in a time context by mapping it to major events terminating that work and corresponding to the conduct of major reviews forming the foundation of the program integrated master schedule (IMS). We see the program plan as a series of nodes arranged in a series-parallel network mapped to trios in our matrix illustrated in Figure 77. These nodes, in turn, are traceable to a combination of contributing functional departments, product architecture or WBS, and work elements.

In that we should form the work teams focused on the product entity structure or WBS, each team inherits a contiguous block of budget in the WBS and work in the corresponding SOW and IMP section, a clearly correlated schedule (IMS expands upon the IMP in the time axis), and a specification at the top level of the hierarchy for which they are responsible. The cross-organizational interfaces, which lead to most program development problems, become exactly correlated with communication patterns that must occur between the teams to develop those interfaces, and thus the people are clearly accountable for doing this work well.

This single line WBS has been discussed in terms of a physical product but most products also require deliverable services such as training, system test and evaluation work, site activation work, and other services. Figure 77 suggests that these services can be framed in the context of physical entities, some of which may be delivered and others not. The training program may entail the services of teaching the courses developed initially or for the life cycle. Indeed, the whole system could be operated by the contractor throughout its life and this would clearly be a service. But, these services can be related to the things in the system. The author distinguishes between the services that would be performed on the delivered system and those required to create the system. The latter should all be coordinated with the product physical product.

Well, how about a contract that entails only services? The WBS for such a contract can still be formed of services to be delivered, but there remains a distinction between the work that must be done to prepare for performing those services and the performance of those services. The former can be oriented around an architecture of services being designed. The latter constitutes the product that is delivered. The proposition is that the services performed to develop the deliverable services should be mapped to the deliverable services for WBS purposes.

There may still be an argument that there are some traditional work elements that will simply not map to the product structure shown in Figure 77, so the elements listed in the appendices of canceled military standard MIL-STD-881A were reviewed for correlation. The traditional services-oriented (nonprime item) elements are listed in Table 12 correlated with the element of the product-only WBS illustrated in Figure 76 to which they would be attached. The author's preferred base 60 identification system has been used for WBS, but the reader may substitute a logistics numbering system, the common four-digit structure, or a decimal delimited style if desired.

Table 12 Non-Product WBS Transform

Non-prime item work element	Comment
Training	A24
Peculiar Support Equipment	A21
Systems Test and Evaluation	A1 elements as appropriate to the test. System testing may be mapped to architecture item A or A1 depending on the scope of the planned testing. Otherwise, the work maps to the activities associated with items within architecture A1.
System Engineering/Program Management	Elements of system A as appropriate. System engineering work accomplished at the system level would map to A. Deliverable operational product element work would map to A1. This same pattern would be extended down to team management and the systems work accomplished within the context of those teams.
Data	A22
Operational Site Activation	A3 as the work relates to facilitization and preparation of sites from which the product will be operated and maintained.
Common Support Equipment	A21
Industrial Facilities	A4, A5
Initial Spares and Initial Repair Parts	A24

In the product-only WBS all of the system services or processes are simply part of the enterprise product development process and they are all linked through the program planning process to product entities where they pick up development budget and schedule. Figure 78 illustrates the overall generic planning to program planning transform that the product-only WBS fully enables.

The enterprise has prepared itself by developing an identity composed of an organizational charter plane showing the map between common tasks that must be applied on all programs and their functional department structure. Each functional department has developed written practices telling how to do the tasks mapped to them coordinated with the designated task lead department's practice and tools available to do the work. This how-to information could be in detailed text format or be provided in training

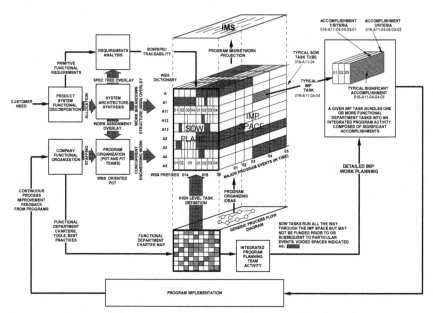

Figure 78 The grand planning environment using product-only WBS.

materials traceable to the content of an abbreviated practice. This generic identity data is continuously improved based on lessons learned from on-going programs and should be the basis for in-house system engineering training.

When a program comes into the house, the work is planned by first performing a functional decomposition and analysis on the customer need, which yields a preferred architecture synthesized from the allocated functionality and the aggregate views of engineering, manufacturing, procurement and material, finance, logistics, and quality within the context of a program integration team (PIT). The product architecture becomes the WBS (base 60 identification used here can be substituted with any other system), which is the basis for development of the program statement of work (SOW). The SOW is created by the PIT by linking high-level tasks with the WBS elements, as discussed earlier. Figure 78 includes WBS prefixes to cover the most complicated case, where one prefix may be for recurring and the other for nonrecurring.

The WBS elements also become the basis for formation of the product development teams (PDT), each of which receives a contiguous block of WBS, SOW, and related budget in a simple fashion that will lead to a need for relatively low management intensity. The continuing requirements analysis process performed by the PIT yields a specification for each team defining the design problem to be solved by the team within the context of the planning data and budget.

The IMP is formed by expansion of the SOW with additional detail that tells what has to be done and by whom it must be done. How-to information is provided to all programs by reference in the IMP to generic how-to practices provided by the functional departments rather than through IMP narratives unique to each program. All IMP tasks at some level on indenture are linked to major program events, through the IMP task coding technique. The events are used to define planned milestones coordinated with major program reviews. Scheduling people craft the raw information into an integrated master schedule (IMS). Teams expand upon the IMP to provide details about their planned activity so as to be consistent with the IMS. This process can be aided by possession of a generic process flow diagram or network that places generic tasks at some level of indenture into a sequence context.

The work definition flows into the IMP by mapping functional department charters to SOW tasks and performing integration and optimization on the details to craft a least-cost, most effective process for developing a product that satisfies the customer need. In the U.S. Air Force integrated management system, these functional tasks mapped into the IMP tasks become the significant accomplishments phrased in terms of the result of having performed that work. Each of these tasks includes at least one accomplishment criteria whereby one can tell when the task is finished.

12.6 FOCUSING ON THE WHOLE

There are so many pieces of a successful verification activity on a program and this book covers many of them. It is not enough for an organization to have people on the payroll who are gifted in the details of a process. It is necessary but not sufficient. The organization must be capable of fusing together the skills of the individuals to form a powerful communications network yielding a tremendous data rate between these individuals. It is through the application of gifted engineers in the right specialized fields on difficult jobs joined together through an effective communication process that we achieve great ends in the development of products to satisfy complex and demanding needs.

The proposition is offered that great strides in the performance of system engineering will seldom come from education alone (another necessary but not sufficient case). The organization that would perform this process well must have a shared corporate memory of product and process knowledge defined in source materials available to all, which acts as a foundation for a tremendously effective conversation between the participants. In the ideal application, the participants retain their individuality but during working hours become elements of the equivalent of one great thinking machine driven by their common knowledge base and effective communications media that, aided by gifted system engineers, integrates their evolving specialized findings into the whole.

These conditions will not materialize from a cloud of stars off your company's fairy godmother's baton tip. They must be brought into being through the result of good management, starting at the top, and must be moved down and throughout the organization by a determined management staff. Some of the principles noted in the prior chapters that encourage a desirable result are:

a. An ultimate enterprise manager who understands the need for both specialization and integration as well as how the systems approach and function fits into this pattern of human behavior. This person must also have the force of personality to singlemindedly emphasize the necessity that everyone cooperate in implementing a documented process as well as contributing to continuous improvement of that process.

b. A matrix structure that requires functional management to deliver to programs good people, good tools, and good practices as well as provide service on program boards and teams in advisory and critical review functions. The matrix structure should also call upon programs to provide the strong day-to-day management function for the people assigned to the programs. These programs should be organized as cross-functional teams oriented about the product architecture.

c. The program teams identify and define requirements as a prerequisite to application of concurrent design principles to develop compliant design solutions. Requirements are reviewed and approved prior to authorizing detail design that, in turn, is reviewed and approved prior to authorizing manufacture of qualification articles. There is a good condition of traceability between the development requirements, verification requirements, and verification plans and procedures leading to well-executed verification work that discloses accurately the condition of compliance of the designs with the requirements.

d. Risk is minimized through a consciously applied program to identify potential problems and find ways to reduce the probability of occurrence and adverse consequences that may result.

The importance of a clear expression of management support for systems work cannot be overemphasized. The author encountered a good example of management support while completing a string of two 96-hour system engineering training programs at Hughes Defense Communications in Fort Wayne, Indiana while writing this book. Prior to the beginning of the first course, a videotape of the president of the company (Magnavox at the time), explaining how important it was that the company perform system engineering work well, was run for each class. At the end of the program a graduation exercise included a talk by the head of engineering, encouraging the engineers to work to apply the principles covered in the program and to come to see him if they ran into difficulty implementing those principles on programs.

Management cannot issue encouragement to the troops and then withdraw to the safe ground. This is like encouraging the people of Hungary to revolt back in the 1950s and then not helping them when the Russians brought in the tanks. The effect on your troops is similar. It is very career limiting for the troops and quickly leads to individual burn-out and group failures in the systems approach. After each failed attempt, it becomes all the more difficult to energize the process another time.

Excellence in system engineering, like so many things involving us humans, is simple but it is not easy. It requires the continuous application of energy, brains, and management strength and inspiration. Managers in companies that would do system engineering well must not depend on the troops to bring forth an excellent systems approach; rather, they must lead the movement to continuously improve, not as a response to some new buzzword but a permanent dedication to excellence no matter the intensity of the noise surrounding them. There will be times when others – subordinates, peers, and superiors – will do their best to avoid supporting your ideas with zeal and may even seek to damage your best efforts while doing their best to appear otherwise.

So, the most fundamental characteristic of a successful system engineering supporter is an inner unshakable strength of position based on the sure knowledge that in order to achieve great and complex engineering goals, one must craft a framework of product infrastructure married to teams of individual human beings, each with the right specialized knowledge (and therefore limitations) welded into powerful engines of composite thought matched to the right resources (money, time, work definition, facilities, machines, and materials) and characterized by the simplest aggregate interface between product teams and product items in terms of work definition, budget, planning data, and degree of alignment between product interfaces and communication patterns that must exist between the teams. This message has to be propagated through the organization with steadiness over time while the systems capability is improved to deliver on its promise.

While a young U.S. Marine sergeant, the author was asked by his wife-to-be, "What do you intend to do with your life?" The author answered, "Work with machines; that's where the complexity lies." The author has long since come to realize that the complexity resides in us humans. The infrastructure useful in the development of systems is much more complex than the product produced by it. So, it should be no wonder that it is seldom done to perfection and so often done poorly. The principles of doing system development work well and organizing the work of people to cooperate efficiently in so doing are publically available now in many books, however, so programs are run poorly by program managers unnecessarily. The counter-argument is that there is no more difficult job in system development than managing people in accordance with a plan, schedule, and budget. The problem is that we humans carry around with us the most complex instrument on earth, called a brain, and we do not yet understand all of its complexity. Further, those brains are interfaced via the worst interface on earth – human communications. So, tomorrow start a campaign to become a better system engineer by recognizing that we individually cannot possibly succeed without the whole population of a program doing their best every day to perform their work, thus forming a powerful team of human beings acting as if directed by a single great mind formed from the aggregate of all of their minds, connected through human communications and managed by a fellow imperfect human being, our program manager, in accordance with a plan. While you are in a good mood, say "good morning" to your program manager and do your best that day to help him or her be successful, for they have the most difficult job on the planet.

APPENDIX A

System Verification Template and Data Item Descriptions

A.1 INTRODUCTION

This appendix provides templates for the verification documents referred to in this book. The documents apply at three product levels in a program: (1) at the system level, (2) at the item or interface level, and (3) at the parts, materials, and processes level. A system will consist of some number of items and interfaces. Each item will consist of many parts and materials and be affected by numerous processes applied at different stages of development. The intent is that every aspect of a system being delivered to a customer will have had it verified that its design and manufacture complies with the content of all of the program specifications and other contract provisions.

Verification work will be applied to the system, its items, its interfaces and its parts, materials, and processes in the form of tasks. Verification planning and management work results in identification of some number of tasks and each task should have developed for it a plan, procedure, and report. The book makes the point that the plan and procedure documents may be combined by including the procedural steps in the plan. The documents of interest in verification work will involve plans, procedures, and reports related to the tasks as well as the overall verification process. Detailed template formats are not provided for parts, materials, and processes verification in this appendix because these documents are so commonly prepared by suppliers that control of their format is often very difficult.

The author maintains that an enterprise should possess a template and a data item description (DID) for each of these documents in the best case but suggests that if an enterprise has no templates it could start by building a set of templates and as time permits thereafter build the set of DID. A template is simply an outline for a document that essentially gives the paragraph titles and numbers organizing the document, while a DID offers instructions on how to complete content in the structure provided by a template for any particular program.

A.2 APPLICATION

Ideally, the enterprise should maintain a complete set of templates and DIDs for a program all stored in a protected environment under configuration control and be prepared to transfer a complete set in an agreed format to any new program for use by program personnel to create required program documentation. This appendix provides only the template set. The complete requirements related set for an enterprise would also include document sets for system architecture and requirements engineering (including all types of specifications), system synthesis, and system management as well.

An enterprise should maintain a set of reviewed and approved templates and DID under configuration control. As improvements are offered through program experience, those improvements should be brought to the attention of the functional department responsible for the document in question for consideration as a formal change that would be subjected to review and approval before being introduced into the library of documents.

A.3 PRESENTATION

The appendix presents each document as a sub appendix A1 through A9 as listed in the table of contents. The book is sold either with paper appendix content, a disk that contains Appendix A, or provides for the download of these documents in Microsoft Word so that an enterprise may import the whole set of documents into a computer and edit them for enterprise preferences to begin their own set of template and DID masters for use on programs. The documents are paginated within their own sub appendices.

A.4 TEMPLATE UPDATE

Any improvements suggested by readers addressed to the author at jeff@jogse.com will be evaluated and be used to update any publisher stored information (if any) that can be downloaded by book purchasers.

System DT&E and Item or Interface: Qualification Task Plan Template

1. Task Plan Information
   - **1.1** System DT&E or Item or Interface Qualification Verification Task Number
   - **1.2** Name of System, Item, or Interface Covered
     - **1.2.1** System, Item, or Interface Functional Identification
     - **1.2.2** System, Item, or Interface Specification Identification
     - **1.2.3** Specification Content to Be Verified in Task
   - **1.3** Verification Method Applied
   - **1.4** Objective of Task
   - **1.5** Task Budget Limits
   - **1.6** Task Schedule Limits
   - **1.7** Task Overview
2. Task Personnel Requirements
   - **2.1** Task Principal Engineer
   - **2.2** Other Personnel
3. Task Material Resources
   - **3.1** Identification of Task Resources
   - **3.2** Preparation of Task Resources
   - **3.3** Post Task Applications of Task Resources
   - **3.4** Post Task Adjustments Needed for Task Resources
4. Task Procedural Steps
   - **4.1** Step 1
     - **4.1.1** Step 1 Setup
     - **4.1.2** Execute Step 1
     - **4.1.3** Collect, Evaluate, and Record Results
   - **4.2** Step 2
     - **4.2.1** Step 2 Setup
     - **4.2.2** Execute Step 2
     - **4.2.3** Collect, Evaluate, and Record Results
   - **4.N** Step N

NOTE

Repeat Step N as needed.

    4.N.1. Step N Setup

    4.N.2. Execute Step N

    4.N.3. Collect, Evaluate, and Record Results

5. Post Task Activities

NOTE

Include Table 1 in Paragraph 1.2.3 of task plan and complete with identified data. Requirement ID is used to identify requirements rather than paragraph numbers to avoid the necessity to edit the plan based on specification changes over the run of the program.

Table 1 Task Requirements Traceability Table

SPECIFICATION SECTION 3 RQMT ID	SPECIFICATION SECTION 4		TASK PLAN STEP	TASK PLAN SECTION 4 PARAGRAPH 4N3
	TITLE	RQMT ID		

APPENDIX A2

System DT&E and Item or Interface: Qualification Task Report Template

1. Introduction
   1.1 Purpose
   1.2 Verification Item Definition

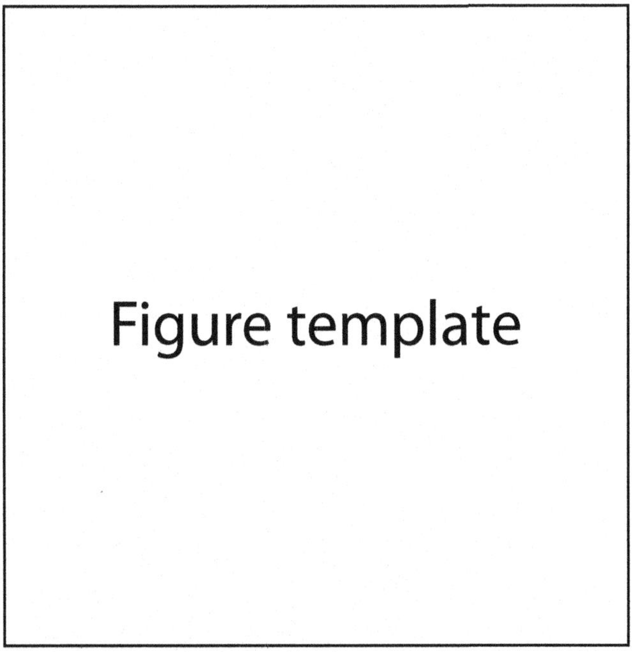

Figure 1 Verification Item Physical View

   1.3 Identification of the Basis for Verification
   1.4 Verification Method Applied
   1.5 Extent of Promotion and Demotion Applied
2. Overview of Task

3. Results of Task
3.1 Task Verification Traceability Table

Table 1 Task Verification Traceability Table

STEP	SPEC REQ ID	TITLE	VER REQ ID	REPORT EVIDENCE LOCATION	Result

3.2 Results Interpretation
3.2.N Line N Information
4. Task Conclusion

Program Functional Configuration Audit: Report Template

1. Introduction
   - **1.1** System Entity Qualified
   - **1.2** Performance Specification Version Applicable
2. Summary of Qualification Task Results
   - **2.1** Summary of Test Qualification Task Results
     - **2.1.1** Test Task 1 Results
     - **2.1.N** Test Task N Results
   - **2.2** Summary of Demonstration Qualification Task Results
     - **2.2.1** Demonstration Task 1 Results
     - **2.2.N** Demonstration Task N Results
   - **2.3** Summary of Analysis Qualification Task Results
     - **2.3.1** Analysis Task 1 Results
     - **2.3.N** Analysis Task N Results
   - **2.4** Summary of Examination Qualification Task Results
     - **2.4.1** Examination Task 1 Results
     - **2.4.N** Examination Task N Results
   - **2.5** Summary of Special Qualification Task Results
     - **2.5.1** Special Task 1 Results
     - **2.5.N** Special Task N Results
3. Degree of Compliance Achieved
4. Adverse Results Observed
   - **4.1** Adverse Result 1
     - **4.1.1** Course of Action Taken
     - **4.1.2** Consequence of Corrective Action
   - **4.N** Adverse Result N
5. Final Degree of Design Compliance Achieved
6. Recommendations

Item or Interface Acceptance Verification: Task Plan Template

1 Task Plan Information
  1.1 Item or Interface Acceptance Verification Task Number
  1.2 Name of Item or Interface Covered
    1.2.1 Item or Interface Functional Identification
    1.2.2 Item or Interface Detail Specification Identification
    1.2.3 Specification Content to Be Verified in Task
  1.3 Verification Method Applied
  1.4 Objective of Task
  1.5 Task Budget Limits
  1.6 Task Schedule Limits
  1.7 Task Overview
2. Task Personnel Requirements
  2.1 Task Principal Engineer
  2.2 Other Personnel
3. Task Material Resources
  3.1 Identification of Task Resources
  3.2 Preparation of Task Resources
  3.3 Post Task Application of Resources
  3.4 Post Task Adjustments Needed for Task Resources
4. Task Procedural Steps
  4.1 Step 1
    4.1.1 Step 1 Setup
    4.1.2 Execute Step 1
    4.1.3 Collect, Evaluate, and Record Results
  4.2 Step 2
    4.2.1 Step 2 Setup
    4.2.2 Execute Step 2
    4.2.3 Collect, Evaluate, and Record Results

NOTE

Repeat Step N as needed.

> **4.N** Step N
> > **4.N.1** Step N Setup
> > **4.N.2** Execute Step N
> > **4.N.3** Collect, Evaluate, and Record Results

NOTE

Include Table 1 in Paragraph 1.2.3 of task plan and complete with identified data. Requirement ID is used to identify requirements rather than paragraph numbers to avoid the necessity to edit the plan based on specification changes over the run of the program.

Table 1 Task Requirements Traceability Table

Task plan Step	Specification section 3 RQMT ID	Title	Specification section 4 RQMT ID	Task plan section 4 Paragraph 4 N3

Item or Interface Acceptance Verification: Task Report Template

APPENDIX A6

Program Physical Configuration Audit: Report Template

1. Introduction
   - 1.1 System Entity Accepted
   - 1.2 Detailed Specification Version Applicable
2. Summary of Acceptance Task Results
   - 2.1 Summary of Test Acceptance Task Results
     - 2.1.1 Test Task 1 Results
     - 2.1.N Test Task N Results
   - 2.2 Summary of Demonstration Acceptance Task Results
     - 2.2.1 Demonstration Task 1 Results
     - 2.2.N Demonstration Task N Results
   - 2.3 Summary of Analysis Acceptance Task Results
     - 2.3.1 Analysis Task 1 Results
     - 2.3.N Analysis Task N Results
   - 2.4 Summary of Examination Acceptance Task Results
     - 2.4.1 Examination Task 1 Results
     - 2.4.N Examination Task N Results
   - 2.5 Summary of Special Acceptance Task Results
     - 2.5.1 Special Task 1 Results
     - 2.5.N Special Task N Results
3. Degree of Compliance Achieved
4. Adverse Results Observed
   - 4.1 Adverse Result 1
     - 4.1.1 Course of Action Taken
     - 4.1.2 Consequence of Corrective Action
   - 4.N Adverse Result N
5. Final Degree of Design Compliance Achieved
6. Recommendations

System Development Test and Evaluation: Master Plan Template

1. INTRODUCTION
   1.1 Purpose
   1.2 Mission Description
   1.3 System Description
      1.3.1 System Threat Assessment
      1.3.2 Program Background
         1.3.2.1 Previous Testing
      1.3.3 Key Capabilities
         1.3.3.1 Key Interfaces
         1.3.3.2 Special Test or Certification Requirements
         1.3.3.3 Systems Engineering (SE) Requirements
2. TEST PROGRAM MANAGEMENT AND SCHEDULE
   2.1 T&E Management
      2.1.1 T&E Organizational Construct
   2.2 Common T&E Data Base Requirements
   2.3 Deficiency Reporting
   2.4 TEMP Updates
   2.5 Integrated Test Program Schedule
3. TEST AND EVALUATION STRATEGY
   3.1 T&E Strategy
   3.2 Evaluation Framework
   3.3 Developmental Evaluation Approach
      3.3.1 Mission-Oriented Approach
      3.3.2 Developmental Test Objectives
      3.3.3 Modeling and Simulation
      3.3.4 Test Limitations
   3.4 Live Fire Evaluation Approach
      3.4.1 Live Fire Test Objectives
      3.4.2 Modeling and Simulation
      3.4.3 Test Limitations

System Development Test and Evaluation: Report Template

1. Report Introduction
2. Task Summaries
  2.1 Test Task Summaries
    2.1.1 Test Task 1
    2.1.N Test Task N
  2.2 Demonstration Task Summaries
    2.2.1 Demonstration Task 1
    2.2.N Demonstration Task N
  2.3 Analysis Task Summaries
    2.3.1 Analysis Task 1
    2.3.N Analysis Task N
  2.4 Examination Task Summaries
    2.4.1 Examination Task 1
    2.4.N Examination Task N
  2.5 Special Task Summaries
    2.5.1 Special Task 1
    2.5.N Special Task N
3. Summary Report of Findings
4. Recommendations

Program Parts, Materials, and Processes: Master Plan Template

1. Introduction
   1.1 Purpose and Objective
   1.2 Safety Issues
   1.3 Security Issues
   1.4 Maintainability Issues
   1.5 Program Financial Issues
2. Parts
   2.1 Parts Approval and Changes to Approved List
   2.2 Software Parts Control
   2.3 Long Lead
3. Materials
   3.1 Materials Approval and Changes to Approved List
   3.2 Long Lead
4. Processes
   4.1 Processes Approval and Changes to Processes List
   4.2 Software Processes Control

NOTE

On a large program the preceding template will probably not work well and should be replaced by a separate template for each kind of entity (part, material, process). In each case a template can be formed with a section 1 and a section 2 with the latter composed of the content of section 2, 3, or 4 above as a function of the kind of entity covered. In this case, drop the word MASTER in the template title.

INDEX

Note: Page numbers followed by *f* indicate figures and *t* indicate tables.

Printed in the United States
By Bookmasters